PREPARING FOR DISASTER

AUTHOR BIOGRAPHIES

Dr Douglas Paton is a Professor in Psychology at the University of Tasmania. He is also a Research Fellow at the Joint Centre for Disaster Research (New Zealand), a Technical Advisor on risk communication to the World Health Organization, a member of the IRDR Risk Interpretation and Action sub-committee of the UN-ISDR, and a Principle Scientific Advisor to the Bushfire Cooperative Research Centre. He has published 18 books, 70 book chapters and 130 peer reviewed articles. Douglas has extensive experience in developing and testing models of community resilience and adaptive capacity for natural hazards and risks communication programs and is currently involved in all-hazards and cross cultural research in Australia (bushfire, flooding, tsunami), New Zealand (earthquake, volcanic), Japan (earthquake, volcanic), Indonesia (volcanic), Taiwan (earthquake, typhoon), and Portugal (bushfire).

Dr John McClure is Professor in Psychology at Victoria University of Wellington. He completed his PhD at The University of Oxford, and his book on Explanations Accounts and Illusions has just been published in paperback form by Cambridge University Press. He has published over 60 peer reviewed research papers, many of which focus on psychological factors that affect preparation for hazards, especially earthquakes. He led research funded by the New Zealand Earthquake Commission on factors affecting different types of preparedness in businesses and households, and is researching judgments about low frequency hazards such as earthquakes. He is currently doing research on the effects of the recent earthquake in Canterbury, New Zealand on risk perception and preparedness.

PREPARING FOR DISASTER

Building Household and Community Capacity

By

DOUGLAS PATON, Ph.D., C.PSYCHOL.

School of Psychology
University of Tasmania
Launceston, Tasmania, Australia

and

JOHN McCLURE, Ph.D.

School of Psychology
Victoria University
Wellington, New Zealand

CHARLES C THOMAS • PUBLISHER, LTD.
Springfield • Illinois • U.S.A.

Published and Distributed Throughout the World by

CHARLES C THOMAS • PUBLISHER, LTD.
2600 South First Street
Springfield, Illinois 62704

© 2013 by CHARLES C THOMAS • PUBLISHER, LTD.

ISBN 978-0-398-08895-8 (hard)
ISBN 978-0-398-08896-5 (paper)
ISBN 978-0-398-08897-2 (ebook)

Library of Congress Catalog Card Number: 2013010824

Printed in the United States of America
MM-R-3

Library of Congress Cataloging-in-Publication Data

Paton, Douglas.
    Preparing for disaster : building household and community capacity /
by Douglas Paton, Ph.D., C.Psychol., School of Psychology, University
of Tasmania, John McClure, Ph.D., School of Psychology, Victoria
University.
      pages cm
  Includes bibliographical references and index.
    ISBN 978-0-398-08895-8 (hard) -- ISBN 978-0-398-08896-5 (pbk.) --
ISBN 978-0-398-08897-2 (ebook)
    1. Emergency management. 2. Preparedness. I. McClure, John. II.
Title.

HV551.2.P38 2013
363.34′72–dc23
                                                            2013010824

PREFACE

Throughout human history, societies have been established and have developed, usually as a result of people's desire to profit from, benefit from, enjoy or utilize the physical, economic, and aesthetic amenities afforded by their natural environment, in areas that increase societal exposure to volcanic, wildfire, storm, flooding, tsunami, and seismic systems. Periodically, however, the activity of these often beneficial natural processes can interact with the human settlements in ways that create hazardous conditions for societies, their members and the institutions and infrastructure that people rely on to sustain normal functioning. When this happens, these natural processes become natural hazards.

When societies and their members find themselves, by accident or design, having to co-exist with natural processes capable of threatening life and livelihood, there is much they can do to protect themselves from the potentially adverse consequences of hazard events. However, despite the evident advantages that being prepared confers on people and communities, and the fact that people are often aware of their risk, research has consistently found that individual, community, and business preparedness levels tend to be low. This book examines why this is so and identifies what can be done to expedite the development of sustained preparedness, at household, community, and societal levels. It does so by emphasizing the need for this aspect of social risk management to be based on engagement principles: how people engage with their natural environment, how they engage with each other, and how people and agencies and businesses engage with each other. An engagement-based approach to hazard preparedness portrays preparedness as a process in which multiple stakeholders (people, scientists, risk management specialists, government agencies, businesses, etc.) share responsibility for societal risk management and play complementary roles in how it develops and how it is sustained over time.

Following a discussion of how people relate to the environmental hazards that they need to prepare for, the book then introduces what being comprehensively prepared to manage the impacts of natural hazards means. An

v

analysis of the nature and extent of people's preparedness is used to frame the progressive discussion of how intra-personal processes, social cognitive theories, and social theories can be used to both understand preparedness behaviors and inform the development of sustained individual, community, societal, and business preparedness.

DOUGLAS PATON
JOHN MCCLURE

CONTENTS

PREPARING FOR DISASTER

Chapter 1

CO-EXISTING WITH A
HAZARDOUS ENVIRONMENT

Civilisation exists by geological consent, subject to change without notice.
Will Durant (1885–1981)

INTRODUCTION

Throughout human history, societies have been established and have
developed in locations which have resulted in their members living in
proximity to natural processes associated with, for example, volcanic, wildfire, storm, flooding, tsunami and seismic systems. Decisions to live and
develop in these locations can be attributed, at least in part, to people's desire
to profit from, benefit from, enjoy or utilize the physical, economic, and aesthetic amenities these natural processes afford people and the societies they
create. For example, seismic activity can create natural harbours and spectacular mountain scenery, river systems provide navigable routes for trade,
and settlements established in volcanic regions often do so to benefit from
the fertile soils found in these areas.

However, periodically, the activity of these natural processes can interact
with the human settlements that have developed in these environs in ways
that create hazardous conditions for societies, their members, and the institutions and infrastructure that people rely on to sustain normal functioning.
Thus decisions made in the past, often with little knowledge of the potential
threat posed by natural environmental processes, can result in people suddenly finding themselves living in harm's way. When this happens, these natural processes become natural hazards.

Thus, as Durant points out, nature can impose change on people and it
can do so suddenly. However, the degree of notice of impending activity and

3

the extent of the imposed change that accompanies the action of natural processes are outcomes that are, to some extent at least, within the realm of human influence. This book is concerned with identifying how this influence can be exercised by people, communities, and societies. Understanding how people can exercise this influence starts with appreciating the difference between natural processes, natural hazards, and disasters.

NATURAL PROCESS, NATURAL HAZARDS, AND DISASTERS

It is important to note that natural processes are not hazards per se, and hazard activity cannot always be equated with disaster. Natural processes become hazards when the levels of their activity reach a level of intensity or persistence that threatens people and what people value and can significantly disrupt or destroy the infrastructure, systems, and institutions that sustain societal functions. When the interaction between human settlements and human-use processes (e.g., utilities, transportation, administration, etc.) and natural processes (e.g., hurricanes, earthquakes, and floods, etc.) reaches a level where lives being lost, infrastructure damaged or destroyed, societal functions rendered inoperative, and survivors' lives thrown into disarray, the action of a natural hazard becomes a disaster.

Extreme levels of natural process (e.g., earthquakes) activity can expose populations and social systems to demands and consequences that fall well outside the realm of normal human experience. This can occur suddenly (e.g., as with earthquakes) or more insidiously over periods of time that can be measured in years or decades (e.g., as with environmental hazards such as salinity or drought). When particularly intense and/or prolonged hazard events do occur, they impact on people, they affect communities, disrupt the societal processes that serve to organize and sustain community capacities and functions. Consequently, in the absence of activities implemented specifically to develop a capability or capacity for continued functioning, normal routines (e.g., that rely on often taken for granted access to power and water, transportation, social services, etc.) will no longer be supported or maintained within areas affected by hazard activity.

The potential for interaction between natural processes and human settlements is ever present, but experience of events that pose a threat (i.e., when natural processes occur at levels that present hazards that communities have to respond to) is periodic and generally infrequent. If the nature of the activity of these natural processes can be understood and their physical, personal, and social consequences identified, it becomes possible to develop risk management strategies that can facilitate the ability of people and societies to co-exist with the potentially hazardous elements in their environment. This can be done by mitigating the risk posed by natural hazards and/or developing the beliefs, resources, procedures, and competencies required to facilitate the ability of people, communities, and societies to anticipate what they might have to contend with and develop the knowledge, attitudes, and behaviors required to ensure they can cope with, adapt to, recover from, and learn from experience of natural hazard events and their consequences.

Mitigation and readiness (or preparedness) strategies afford societies and their members several ways in which they can protect themselves from hazard consequences, minimize the harm and adverse consequences they could experience, and enhance their ability to deal with any consequences they do experience. These activities fall under the general heading of risk management. The starting point for the development and implementation of risk management strategies is understanding what has to be mitigated or prepared for.

WHAT DO PEOPLE AND SOCIETIES HAVE TO CONFRONT?

Risk management starts with identifying the natural processes that exist within an environment and developing an understanding of how the action of these processes can create adverse circumstances for people, societies, and the physical and built environments they inhabit. Armed with this knowledge, scientists and risk management specialists are able to identify actions that can be taken to mitigate and/or manage the threat posed to people and communities. This process would be challenging enough if societies only had to contend with a single natural process. This is, however, rarely the case.

Many societies have to contend with their being susceptible to experiencing multiple hazards. For example, as the tragic events in Japan in March 2011 demonstrated, areas prone to offshore earthquakes can be susceptible to experiencing local-sourced tsunami that can strike in minutes. The Pacific Northwest of the USA faces similar risks. Residents in California can experience both geological (e.g., seismic, volcanic) and environmental (e.g., wildfire) hazards. Residents in countries like Taiwan are susceptible to experiencing geological (e.g., earthquake, landslide) and meteorological (e.g., typhoon) hazards. It is, however, possible to identify the hazards that can occur in a given area. The development of an inventory of the range of potentially hazardous natural processes in a given area defines its "hazard-scape."

Hazard-Scapes

The hazard-scape is a compendium of the natural processes from which the hazards a society and its members will have to contend with emanate. The process of populating the hazard-scape provides the foundation for identifying the hazard characteristics (e.g., ground shaking) and behavior (e.g., intensity, duration, distribution) that provide the raw material for risk management. Analyses of hazard characteristics and behavior define both what people and societies will experience (e.g., ground shaking) and how bad (e.g., how intense) these experiences could be. A further challenge to risk management arises from the fact that the hazard-scape is not necessarily a static entity.

The analysis of hazard characteristics and behaviors need not only be restricted to what is known to exist in the present (based on historical analyses). New scientific information about hazards or changes in environmental attributes (e.g., as a result of climate change) can change the hazard-scape in which people live their lives.

Changes to the hazard-scape could occur as a result of, for example, discovering previously unknown fault lines or as a result of seismic activity being triggered by isostatic processes triggered by loss of large ice fields (McGuire, 2012). Another interesting example, from the point of view of it coming out of left field, is recent research on the potential of a giant slab of rock, the so-called Noggin Block, situated near Australia's Great Barrier Reef to collapse and trigger a tsunami (Cairns Post, 2012). If this occurred, it would create a risk that would not previously have enjoyed a high profile in the areas that could be affected. The emergence of new or more severe problems will also arise from the insidious effects of climate change.

Climate change processes are likely to affect both the distribution and intensity of weather and meteorological hazards (e.g., increased hurricane risk, expansion of drought affected areas, more intense wildfires) and so change what people may have to contend with in future. Places exposed to wildfire hazards, such as California, Australia, Portugal, and Chile can expect to experience more frequent, more prolonged and more intense wildfire hazard events (Paton & Tedim, 2012). Climate change may result in areas which have previously enjoyed relatively benign relationships with their environment experiencing risk from new sources (e.g., increased risk of flooding or drought). Furthermore, the beliefs, decisions, and actions of those living with hazardous circumstances introduce another dynamic influence on how a hazard-scape might evolve over time. For example, people's decisions about land use (e.g., farming, land clearance, irrigation, and industrial development) are increasing the levels of acute and chronic environmental degradation (e.g., pollution, drought, salinity, dust storms) people have to cope with.

Continuing population growth and economic and infrastructure development in areas susceptible to experiencing hazards is increasing the potential magnitude and significance of the personal, community, and societal loss and disruption that can result from hazard activity. It does so because growth and development result in more people living in harm's way, and increases the potential loss of life and infrastructure that could occur for a given level of hazard activity. Furthermore, the decisions people make regarding the level of mitigation they are willing to support or the degree to which they ready themselves to deal with hazard consequences introduce additional dynamics qualities to the relationship between hazard-scapes and societies.

One thing is clear. The natural processes from which hazards emanate are enduring aspects of people's lives. Given that hazards are not just going to go away, an important goal of risk management is to develop ways in which people and societies can co-exist with the potentially hazardous circumstances they could experience.

CO-EXISTING WITH A HAZARDOUS ENVIRONMENT

That people can develop strategies to facilitate a capacity for their co-existence with natural hazard activity is evident from observation of communities that face regular exposure to hazard activity. An example of this is Kagoshima in Japan. Due to the town's proximity to Sakurjima volcano, Kagoshima receives ash fall and ballistic debris approximately 113 days per year. In response to this demand, Kagoshima has developed, for example, building codes, ash removal practices, and community attitudes and preparedness to facilitate continuity of societal functions during periodic volcanic episodes.

The activities undertaken in Kagoshima illustrates how when a need to confront hazard consequences prevails, adaptive mechanisms capable of ensuring sustained societal functioning during periods of hazard activity can be established within the fabric of a society. In the case of Kagoshima, the development of these strategies is justified and accepted by its inhabitants as a result of the frequency with which they experience hazard events and consequences. A relatively high level of hazard experience provides citizens with regular reminders of both why being prepared is important and how mitigation and preparedness activities help them maintain normal functioning in the face of hazard activity. However, this is not representative of the circumstances in which most people experience their relationship with natural hazards.

The majority of societies and communities susceptible to experiencing hazard activity face considerably less frequent activity. For hazards with long return periods (see bleow), and thus extended periods of hazard quiescence, people have little opportunity to gain comprehensive experience or knowledge of the hazard phenomena they may have to contend with. A consequence of the infrequent nature of damaging hazard events is that societal development has often taken place in the absence of a realistic understanding of either the potentially hazardous circumstances prevailing in that location or their implications for societal functioning.

While infrequent hazards activity is beneficial from the perspective of societal development, it creates a more challenging set of circumstances for those with responsibility for natural hazard risk management. This is particularly so with regard to those aspects of risk management that call for changes in people's beliefs and behaviors. Making a significant contribution to this challenge is the uncertainty that surrounds natural hazard activity. In order to develop mitigation and readiness strategies and facilitate their sustained adoption, it is important to explore the sources of this uncertainty and how scientists and risk management experts attempt to impose some semblance of meaning on this uncertainty.

NATURAL HAZARDS: SOURCES OF UNCERTAINTY

Current knowledge about natural hazards reflects the outcomes of scientific endeavour going back several centuries. However, a period of systematic research of only a few hundred years is too short, relative to geological history measured in millions of years, to have provided the scientific community with insights into the full range and peak magnitudes, intensities, or durations of hazard activity. However, the analysis of historical experience provides the foundation for imposing structure and meaning on this uncertainty. It does so by identifying hazard characteristics and behaviors (see below). For the scientists who conduct hazard analyses, this is the primary means of their imposing meaning on uncertain hazard activity. For risk management specialists, this in only one input into this process.

Risk managers also have to deal with uncertainty present in the social and built environment. Risk managers have to accommodate issues such as how the development of the built environment, population growth, demographic diversity, and diversity in people's goals and aspirations (that change over time) influence risk. A hazard of a given intensity could have greater or lesser implications for a society depending on societal characteristics at the time of impact.

It is also important to note that scientific knowledge is constantly developing. Consequently, the estimates of risk derived from the various permutations of hazard-society interactions are further complicated by changes in the scientific understanding of hazard processes. Risk management thus draws on knowledge of scientific, development, social, and psychological processes in order to understand and then manage risk in a climate of uncertainty.

This means that risk management needs to combine the retrospective analyses of what has happened in order to understand hazard behavior and characteristics with a prospective focus that allows risk management to anticipate how hazards will interact with dynamic social and physical environments in the future. This is particularly important when making decisions about mitigation. Decisions about structural mitigation (see Chapter 2) have to accommodate changes in development and growth to ensure that they can offer protection to current and future generations. This can involve anticipating future changes in hazard-societal interactions. For example, siting levees and levee size must consider future development in areas to be protected (e.g., development can change river flow patterns, increase flooding risk, and future economic and domestic development can create future risk). This makes understanding hazard-scapes and their implications for risk management a dynamic and evolving enterprise. The risk management process starts with understanding the natural sources of hazardous circumstances.

Data from the scientific analyses of natural processes define the sources of risk faced by societies and their members and informs the development of the planned structural and social measures developed to manage this risk. From these, data hazard characteristics and behaviors can be identified. Hazard characteristics describe these facets of natural processes that cause loss, damage, and harm to people (e.g., ground shaking). These characteristics are present every time that hazard activity occurs. Hazard behaviors, which include facets such as return and precursory periods, speed of onset, intensity, and duration, are variable and can differ substantially from one hazard event to the next.

NATURAL HAZARDS: CHARACTERISTICS AND BEHAVIORS

The first stage in imposing meaning on the natural processes that can threaten people and societies is to identify their characteristics and behaviors. Hazard characteristics (e.g., the ground shaking associated with seismic activity, ash fall from volcanic eruptions, etc.) define what societies and their members have to prepare for. The analysis of hazard behavior, on the other hand, is undertaken to estimate the parameters of hazard characteristics that could prevail within a given hazard-scape (e.g., the intensity of ground shaking that could occur). This section discusses these behaviors. It is important to note that while it is possible to identify and quantify hazard behaviors, these analyses cannot predict what will happen in the future. Rather they estimate the range of behaviors what might happen in future occurrences.

It is the fact that future projections are estimates that move the risk management process from one of relative objectivity (e.g., when it is purely the subject of scientific research) to one that is open to interpretation (particularly when it enters the public domain). Estimates can be interpreted in different ways by scientists, risk managers, and members of the public (see Chapter 2). In this section, the discussion of hazard behaviors is also used to introduce how the uncertainty surrounding future hazard behaviors is open to interpretation by citizens. In so doing, it provides a springboard for subsequent chapters that explore how and why people interpret hazardous circumstances as they do and what this means for risk management. Discussion commences with the hazard behavior that typically receives the most attention: frequency of occurrence.

Frequency

The potential for interaction between natural processes and human settlement is ever-present, but hazardousness is periodic. The frequency of a natural process describes how often it occurs within a specific period of time. Frequency data thus allow approximations of return periods, and thus estimation of when a future hazard event is likely to occur. From a practical perspective, frequency data affects how people make judgments about the urgency of acting. While the raw data used to describe return periods are derived from relatively objective analyses of historical records of past events, people's interpretation of these data introduces a more subjective quality into the process.

A common source of public misunderstanding arises from the widely used practice of deriving frequency estimates by dividing the total number of years over which events have occurred by the total number of events. For example, if an event occurred 10 times in 200 years, it would have a frequency of about 0.05 (i.e., the probability of occurrence in any one year is 0.05 or 5%) or on average one event (e.g., an earthquake) every 20 years. Framing frequency information in this way can influence how people interpret their risk and how they develop their beliefs regarding the urgency or need for action. This is commonly referred to as a 20-year event; and herein lies the problem.

Of particular concern in this context is a tendency for people to interpret one in 20 years as prescribing a specific time frame (e.g., 20 years hence) rather than an annualized probability (i.e., a one in 20 or 5% chance in any given year). Describing frequency in this way fails to capture the inherent variability in frequency of occurrence and can be misleading for non-expert recipients of frequency information. Interpreting this information as signalling an event that occurs every 20 years can reduce the perceived degree of urgency in people's decision making and therefore reducing the likelihood of people preparing for or supporting investment in structural mitigation to reduce future risk. It has been further identified that the longer this time frame, the less likely people are to take action (Paton, Smith, & Johnston, 2005). While this is a complex issue to overcome, recent work on message framing (McClure & Sibley, 2011; McClure, White, & Sibley, 2009) could provide useful guidelines to assist the framing of information in ways that increase the likelihood of people developing realistic estimates and preparing.

Because of the role that frequency data play in defining the perceived urgency in people's risk management decisions, frequency is the characteristic that people generally pay most attention to. Knowing the frequency, however, does not say anything about how damaging or persistent a given event may be when it does occur. Defining the consequences that hazard activity will create for people, and thus what they will have to contend with, requires considering how frequency (i.e., likelihood of occurrence) interacts with hazard magnitude and intensity (that influence hazard consequences) to affect physical risk. This provides the foundation for understanding how the risk people face reflects the product of likelihood estimates and the consequences people have to contend with.

Magnitude

The introduction of magnitude and intensity parameters into the risk management process identifies new sources of uncertainty. With this uncertainty comes more scope for interpreting risk. While frequency can be considered purely as a characteristic of natural processes, magnitude and intensity, as predictive of the consequences that people will face, introduce a need to actively consider how hazards interact with human use systems, particularly buildings and infrastructure (i.e., the greater the magnitude, the greater the damage that will occur). For example, the 2011 earthquake in Canterbury, New Zealand had very different implications from that in 2010 because the former occurred in the city.

Magnitude and intensity parameters inform decisions such as where to build or what building codes (see Chapter 2) are required if people are to increase their capacity to co-exist with hazards. Thus how people use magnitude and intensity information has implications for whether and to what extent a given hazard event will create a disaster (e.g., the higher the earthquake magnitude covered by building codes, the greater will be the capacity of the built environment to withstand shaking effects). Hazard activity is less likely to create a disaster if mitigation choices (e.g., building codes, levee heights) are made to cover the consequences of events that fall at the higher end of the magnitude/intensity spectrum.

The size of a hazard event is described in terms of magnitude estimates of the action of the physical processes that create hazard consequences. Magnitude is measured using some form of energy scale with the magnitude representing energy over the full duration of the event (Gregg & Houghton, 2006). The Richter and Seismic Moment earthquake magnitude scales are examples of instruments that measure the amount of energy released by a given earthquake. For other hazards, such as volcanic eruptions, estimates of the magnitude are described in terms of the volume or mass of erupted products.

Uncertainty is introduced into the process of estimating risk by the fact that hazard events do not occur with constant magnitude. It is not possible to predict the magnitude of a given event. The magnitude of hazard activity from the same physical source (e.g., the same fault line) can differ substantially from one occasion to another. Furthermore, the reason why development occurs in such highly seismic locations as San Francisco, USA, and Wellington, New Zealand reflects the infrequent nature of damaging hazard events and the fact that people are more likely to have experienced events at the lesser end of the magnitude spectrum. This means that people rarely have experience of the full range of possible magnitudes that could occur (and result in their being more likely to make choices that reflect the magnitudes they are more familiar with).

The tendency for people to experience events at the lower end of the magnitude spectrum can influence their beliefs about hazard events in ways that reduce the likelihood of their preparing to deal with hazard effects, result in their underestimating future risk, and that increases the likelihood of their assuming that the magnitudes they have experienced are representative of all possible occurrences (Mileti & O'Brien, 1993; Paton, Millar, & Johnston, 2001). This bias results in people extrapolating from their ability to deal with low-level (but generally, more frequently occurring) impacts a belief in being able to deal with any future event. Because they fail to consider events of greater magnitude, their perceived need to do anything to manage their immediate risk is reduced and their future risk status increased (Paton et al., 2001).

When communicating magnitude information, it is important that people receive information on the range of magnitudes that could occur. In particular, should a low-magnitude event occur, subsequent risk communication should advise of the potential for more extreme circumstances and the need for planning and preparedness to accommodate these possibilities (Paton et al., 2001; Paton & Wright, 2008). In addition to magnitude, it is important to consider the intensity of the hazard events that could impact a given area.

Intensity

Intensity data informs understanding the activity of natural processes in two ways. It provides a direct measure of rate (e.g., as energy per unit time) and indirectly informs developing estimates of likely damage (Gregg & Houghton, 2006). The modified Mercalli earthquake scale and the Fujita tornado scale, for example, include building damage in their estimates of intensity. In contrast, the intensity scale for volcanic eruptions is based on the mass eruption rate in kg/second. With regard to the relationship between intensity measures and damage, intensity scales also accommodate the potential moderating effects of social and engineering factors (e.g., the presence or absence of steel reinforced concrete, shear walls, hurricane clips, etc.) as well as hazard magnitude (Gregg & Houghton, 2006).

How magnitude and intensity data are communicated to the public is an issue requiring careful consideration. Intensity information plays a significant role in the cost-benefit decisions that determine the level of engineering solution developed as part of a mitigation strategy (e.g., a decision is made to mitigate events up to a specific intensity rather than all possible intensities). While planners and risk managers may be aware of the trade-offs involved in selecting the level of intensity that informs the design limitations of a structural mitigation measure (i.e., what a measure such as a levee can contain or withstand before failing), similar objectivity may not enter citizens' thinking. If citizens, in general, are unaware of the relationship between the range of intensities that could occur and the design limits of an engineering solution, they may overestimate the effectiveness of engineering solutions and reduce their own preparedness (which would otherwise serve to reduce the risk). This process, known as risk compensation, is discussed further in Chapter 4. The interpretation of magnitude and intensity information on decision making is further complicated by the need to consider how they are distributed over time. That is, to consider the duration of the consequences people may have to contend with.

Duration

Natural hazard events can vary in duration from seconds, minutes, and hours (e.g., earthquakes, tornadoes, landslides, avalanches, and flash floods) to days/weeks (e.g., some floods, wildfires) to decades (e.g., some volcanic eruptions, drought, climate change phenomena, and soil salinity). Duration need not always be continuous. Earthquake duration can be experienced as a series of acute events but spread over time, as areas are affected by after-shocks that can persist for several months. This was a characteristic of the aftershock behavior in Christchurch, New Zealand in late 2010 and that con-tinued through 2011 and into 2012. Indeed, an intense aftershock had tragic consequences for the residents of Christchurch some six months after the ini-tial earthquake. It can, however, be difficult to convey to people the full im-plications of an extended aftershock sequence (e.g., the possibility of pro-gressive damage, repeated need to be self-reliant). Even if not causing dam-age, prolonged experience of aftershocks can take a substantial (particularly psychological) toll on those experiencing them (see Chapter 3).

Knowledge of the duration of hazard activity and how and why it occurs plays a significant role in natural hazard planning. It has a significant impact on the time frame that societal systems may be disrupted and thus the peri-od of time during which people may be called upon to adapt to circum-stances in which they will be without or have limited access to normal soci-etal services and functions. As the context of hazard activity progresses, from the shorter to the longer end of the duration spectrum, planning and com-munication processes become increasingly complex.

Accommodating duration estimates in risk management planning also introduces a need to consider secondary hazards. For example, volcanic eruptions at Mount Pinatubo in the Philippines created direct and indirect damage on communities, crops and infrastructure. Communities situated on the flanks of Mount Pinatubo in the Philippines had to contend repeatedly with the lahar (volcanic mud flow) hazard associated with annual rain-in-duced remobilization of unconsolidated pyroclastic material for several years following the eruptions in 1991 (Newhall & Punongbayan, 1996). Secondary consequences from loss of utilities (e.g., power, water, sewerage) after an earthquake can result in an extended period of disruption and present unique management and adaptive pressures for affected populations.

It is also important to appreciate how interaction between factors such as duration and intensity influence hazard-scape characteristics. For example, long duration, high-intensity events present more threatening contexts for coping and adaptation. However, the implications of the interaction between duration and intensity tend not to feature highly in people's decisions about mitigation and preparedness and yet information on how these parameters interact provides people with information on how long they may have to be prepared to meet their own needs without access to normal societal resources and functions.

Duration has other implications for risk management. Gregg and Houghton (2006) pointed out that events that peak early may hamper response and recovery efforts through prolonged direct and secondary impacts. In contrast, events whose intensity is spread over a longer period of time require correspondingly longer periods of response and response management (e.g., equipment to remove volcanic ash, areas to store removed ash) activity. This can extend the period of societal disruption. This, in turn, requires societal planning and readiness planning to accommodate extended and possibly intermittent periods of disruption to and/or loss of services and functions over prolonged periods of time.

The latter example described time frames measured post-event. Time is a variable that also has implications in a pre-event context. This relates to whether or not a hazard affords opportunities to warn people of an impending event.

Precursory Period and Response Time

A distinction can be drawn between the potential lead time a hazard may provide through precursory activity and the time required for the hazard to adversely affect an area once it has developed (Gregg & Houghton, 2006). The term *Precursory Period* (often referred to as "warning time") refers to the time interval between detection of precursory activity and the onset of hazard activity. An example of precursory period is the time delay between the beginning of natural tremor preceding a volcanic eruption (i.e., a type of high-frequency, low-amplitude seismicity associated with magma moving underground) and the surface eruption of the magma. Precursory periods can range from minutes (locally-generated tsunami) to months or years for some volcanic events.

In contrast, the term *Response Time* (also referred to as "speed of onset") refers to the time interval between the beginning or first detection of the hazard (e.g., the formation of tsunami, or detection of eruption activity) and that time when the hazard activity begins to impact adversely upon an area (e.g., when a tsunami reaches a coastal community or lava inundates a settlement). This period can vary from minutes to weeks or longer. Response time is important because it defines the maximum time people have to ready themselves to respond. People's ability to respond, rather than their being forced to react, is a function of their being prepared well before a hazard event occurs. However, problems can arise if people interpret the warning as a signal to prepare rather than as a cue to activate their existing preparedness (see Chapter 3).

It is desirable to distinguish between the potential lead time a hazard may provide through precursory activity and the time required for the hazard to adversely impact an area once the hazard has developed (response time). This can be highly variable. For example, the time between detection of activity and the onset of activity and the subsequent speed of events has varied considerably from a single volcano. Lead times from Mauna Loa volcano (Hawai'i) have ranged from 20 minutes (in 1950) to 115 minutes (in 1984). The time taken for lava to reach the ocean from the same volcano has ranged from 3.5 hours (in 1950) to eight days (in 1926).

These examples illustrate how people can be presented with a range of possible warning and lead times and this may affect the judgments they make about, for example, the time they are likely to have to respond. If people focus on the 3.5-hour data, they could be more likely to prepare in advance for such eventualities. If, on the other hand, they focus on the eight-day data, they may be more likely to believe that no action on their part is required until hazard activity is detected. The latter leaves people more vulnerable if the 3.5 hour scenario was what eventuated in the next eruption. There is no way of knowing beforehand which of these various scenarios could prevail. However, it people have a choice, there is a good chance they will select the option that is least threatening to them or that affords them the greatest latitude in what they should do and when (see also Chapters 3 and 4).

The examples presented in the preceding paragraphs illustrate how people's interpretation of a range of precursory period and response values (that reflect historical variability) needs to be accommodated in risk management planning. This issue is at its most significant for hazards that offer no (e.g., earthquakes) or very limited warning (e.g., local-source tsunami). Under these circumstances, people must be prepared in advance as even the most effective warning process will leave no time for people to implement appropriate safety measures. In contrast, some volcanic events may be accompanied by precursory activity that spans weeks, months, or years, but the actual timing of an eruption can remain difficult to predict. For the latter, the relationship between hazard and society poses a very different kind of challenge. While providing ample time to prepare, the problem may be sustaining people's preparedness if no hazard activity occurs.

Long precursory periods can have other consequences. For example, the long precursory periods that accompany some volcanic eruptions could create considerable economic fallout from withdrawal of capital and business investment and business activity from areas threatened by an impending eruption. The fact that this is the case is one way of introducing a need to consider the spatial distribution of their activity and consequences.

Spatial Distribution

On a spatial scale, hazard activity and its consequences may not coincide. For example, for a large earthquake, ground shaking may be felt over wide regions, but the distribution and extent of damage may be limited either to areas close to the epicentre and/or areas where geologic (e.g., type of rock strata) conditions amplify (or attenuate) the damaging effects of shaking (Gregg & Houghton, 2006). The distribution of consequences is also influenced by the adoption (or non-adoption) of structural mitigation measures (e.g., building design, levels of retrofitting to safeguard against damage from ground shaking). However, for some other hazards, knowing the location of the source may provide few insights into the distribution of consequences. For example, volcanic hazards (e.g., pyroclastic density currents, lava flows, and ballistic blocks) can create consequences covering areas on a scale of kilometres to tens of kilometres, with large volcanic eruptions that produce abundant gas and fine ash in high plumes creating consequences that have regional to global consequences (Mills, 2000). Several hundred thousand people can testify to this as a result of the disruption to air travel that accompanied the eruption of Eyjafjallajökull volcano in Iceland in 2010.

The spatial distribution of hazard consequences and people's beliefs about this aspect of hazard behavior have other implications for risk management. Support for mitigation is less likely to be forthcoming from those who do not believe they could be affected (e.g., people living hundreds of kilometres from a volcano). Even if not directly threatened, those living at some distance from a volcano could be affected directly (e.g., wind-blown ash) and indirectly (e.g., economic fall-out from eruption consequences).

The distribution of volcanic and wildfire hazards can be affected by how they interact with dynamic meteorological factors such as rain and wind (e.g., strength and direction). The distribution of hazards, such as smoke from wildfires or volcanic ash, can have spatial distributions that vary over time and have different intensities. For example, volcanic ash will have less impact on buildings when dry compared with when it is wet from rainfall (which increases weight and acidic effects). Thus meteorological conditions can introduce a dynamic quality into hazard distribution and result in their impacting people who may live tens if not hundreds of kilometres away from the sources of the hazard.

Understanding how to influence support for mitigation and readiness in areas removed from, but susceptible to, hazard consequences is an important area for future research. Meteorological events (e.g., high rainfall, high winds) are an obvious example of hazards whose spatial distribution can change over time. Hazards can also be differentiated with regard to the time of year in which they typically occur.

Temporal Distribution

Hazards associated with meteorological and hydrological processes are, by their nature, more likely to occur at certain times of the year. Hazards that fall into this category include tropical cyclones/hurricanes/typhoons, tornadoes, floods, and wildfire. For example, wildfire is more likely in the hotter summer months than in mid-winter. People are consequently less likely to be interested in information about this hazard during the winter months, yet it is often in late winter and early spring that they should be thinking about preparing.

Temporal issues can also present on a longer time scale. The El Nino Southern Oscillation occurs on a scale measured in years (Gregg & Houghton, 2006). While seasonality may not ordinarily be linked to influencing the occurrence of geological hazards such as volcanic eruptions and earthquakes, this may not always be the case. For example, Cochran, Vidale, and Tanak (2004) found a correlation between the occurrence of shallow thrust faults and the occurrence of the strongest tides, with activity being influenced by changes on the weight of water pressing on a seismically active area. McGuire (2012) discusses a similar process that could occur following loss of ice sheets in higher latitudes, with isostatic processes increasing the potential for earthquakes and tsunami hazards over time.

This brief overview of hazard behavior has introduced the diverse sources of uncertainty implicit within any hazard-scape. It also describes the context in which mitigation and readiness planning takes place and in which risk management strategies are developed and implemented. The process of developing risk management strategies also results in a wide range of stakeholders becoming involved in the process and introduces the need to consider the implications of risk management for people and societies.

SOCIETAL IMPLICATIONS

Knowledge of hazard behavior results from relatively objective scientific analyses of hazard activity. However, while it is possible to say that hazard events will occur, it is difficult, if not impossible, to specify in advance the actual behavior (e.g., intensity, duration) that will accompany a specific event. The consequent uncertainty that is implicit in the hazard-scape in which people are called upon to make their risk management choices means that there is considerable scope for planners and citizens alike to interpret hazard information and its implications in different ways. Furthermore, scientists, risk managers, and citizens differ in the training and interpretive capabilities they bring to the process of making sense of the uncertainties that surround natural hazard activity and that they encounter when becoming involved in risk management (both as suppliers of risk information and as recipients of it).

This book focuses primarily on the risk management issues that arise in the course of attempting to facilitate hazard preparedness in citizens living in areas susceptible to experiencing significant hazard consequences. It discusses the problems that arise in this context, and the solutions that can be implemented to deal with them, with regard to the interpretive opportunities that arise as a result of the implicit uncertainty surrounding natural hazard activity.

The interpretive biases and processes that people introduce into the risk management practice are discussed in several ways. This chapter introduced how people's construal of hazard behaviors (e.g., frequency, intensity) can influence their interpretation of their risk as well as what they do to manage their risk and what activities they may be willing to support in order to manage their risk (e.g., the level of support for mitigation measures—see Chapter 2). This chapter and Chapter 2 essentially describe the context in which people engage in the risk management process.

A book about hazard preparedness clearly needs to define what preparedness is. Chapter 3 introduces the specific activities people can perform or implement to enhance their hazard preparedness. It also discusses how people's interpretation of these various activities exercises an influence on what people do or don't do to prepare, as well as when they are likely to do so. Then Chapters 4 to 7 progressively set forth the intra-personal, social-cognitive, and social and relationship influences on people's interpretation of hazards and preparedness.

While the central focus is on the citizenry who are the recipients of much risk communication, it is important to view them not just as recipients but as partners in the risk management process. Chapter 2 introduces how differences between stakeholders with regard to how they make sense of both environmental hazards and information about them makes it important to appreciate the need for risk communication, public education, and community outreach to be conceptualized in terms of the interdependencies that exist between stakeholders. For example, citizens need information that meets their needs from risk management sources and risk management sources need people to share responsibility for risk management and to act on the recommendations they present. This process of engagement and making complementary contributions to risk management starts with knowing the hazards, but that is only the start.

The task of conducting hazard analyses, and collating and disseminating the findings is normally the sole preserve of earth scientists. However, when it enters the public domain, the meaning attributed to the information from hazard analyses and the development and implementation of risk management goals becomes increasingly open to being influenced by the diverse psychological, social, economic, cultural, and political perspectives that each stakeholder brings to the process of risk management.

This last point introduces the need for risk management issues to be analyzed at several levels of analysis: people, the communities in which they are members, risk managers, politicians, business owners, and so on. Each level of analysis brings different but equally valid and legitimate perspectives on risk management and this, in turn, influences how they engage with the risk management process. They all have stakes in how risk management strategies develop, whether they should or need to be implemented, how they should be implemented and by whom, and how the costs and benefits of doing so should be distributed.

Recognition of the importance of these issues makes it important to conceptualize risk management as a process that involves the input from diverse stakeholders and one in which public involvement, community engagement, and consultation play prominent roles in developing and implementing hazard mitigation and readiness strategies. To achieve this kind of integration, it is essential to understand the interpretive processes used by each group and to accommodate these in the development and implementation of risk management strategies. This is picked up in Chapter 2, which uses the topic of mitigation to introduce the need for risk management to be conceptualized as a process in which diverse stakeholders need to be engaged in ways that facilitate their ability to make complementary contributions to risk management.

Chapter 2

PEOPLE, HAZARDS, AND
HAZARD MITIGATION

INTRODUCTION

When societies and their members find themselves, by accident or design, having to co-exist with natural processes (e.g., seismic, volcanic) capable of threatening life and livelihood, there is much they can do to protect themselves from the potentially adverse consequences of hazard events. They can take steps to mitigate their risk and/or develop their readiness to respond to what could occur at some (uncertain) time in the future. Strategies for facilitating these outcomes are built on a foundation of understanding how the characteristics and behaviors of natural processes (see Chapter 1) interact with people and human use systems to create hazardous circumstances.

Knowledge of hazard characteristics clarifies what people will have to contend with (e.g., ground shaking accompanying earthquakes, ash fall on roofs after volcanic eruptions) should a hazard event occur. Knowledge of hazard behavior informs understanding of, and the quantification of, how likely it is that a hazard event will occur (estimated using frequency or return period data) and the social, environmental, and ecological consequences that could occur (based on identifying a range of possible intensity, magnitude, and duration values based on historical analysis—see Chapter 1). The question then becomes one of how to use knowledge of hazard characteristics and behavior to inform the development and application of strategies to manage risk and develop resilience and adaptive capacity in people and societies.

This task commences when risk management professionals interpret the scientific analysis of the behavior and characteristics of natural processes to make judgments about the degree to which these processes pose a threat to a society and its members (in a given area) and to what people value and

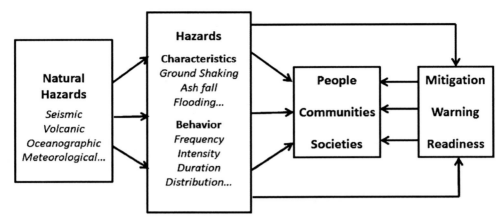

Figure 2.1. Summary of relationship between hazards, people, and risk management strategies.

depend on to sustain everyday life. If the assessment of this risk (the product of estimates of likelihood and consequences) reaches a level deemed unacceptable (according to criteria that are informed by both knowledge of potential hazard consequences *and* political, economic and social concerns), societies can implement several actions to mitigate this risk using the risk management process.

RISK MANAGEMENT

Risk management describes the assessment, planning, and intervention processes and activities that underpin the choices societies and their members (individually and collectively) make about mitigating risk and facilitating their readiness to respond to hazard events. These relationships are summarized in Figure 2.1. From a practical perspective, the scientific analysis of hazard behavior and characteristics informs the development of three general risk management activities: mitigation, warnings, and readiness/preparedness.

The term mitigation encapsulates a suite of strategies designed to reduce societal susceptibility to loss by implementing activities in advance of hazard events to prevent or minimize the potential for a hazard event to become a disaster (Godschalk, Beatley, Berke, Brower, & Edward, 1999). Mitigation encompasses one or more of several activities. It includes strategies such as land-use planning to protect areas susceptible to hazard activity and/or constructing engineered structures that can withstand or divert specific hazard consequences. Mitigation can also include developing building codes and standards that facilitate the capacity of residential, commercial, and public buildings to withstand the action of hazards (at least up to a point).

Acknowledgement of the existence of hazards whose activity cannot be prevented and/or whose activity could occur at levels that exceed the design parameters incorporated in mitigation measures results in the inclusion of warning and readiness strategies in risk management planning. This second group of strategies focuses on monitoring activity and issuing warnings if potentially dangerous hazard conditions are detected or anticipated. The final set of risk management strategies is concerned with preparing people to respond. Strategies in this group focus on facilitating personal, community, and institutional (e.g., work places) readiness in ways intended to increase citizens' safety and their capacity to respond in effective and timely ways to hazard events.

Mitigation and warning systems can be developed relatively independently of the people they are intended to protect. However, the fact that the choices made may reflect the outcomes of democratic processes (e.g., voting for or against mitigation measures) makes understanding how people relate to hazard issues an important consideration in mitigation planning. A strong case can be made for involving or engaging people in the development and implementation of mitigation plans and warning systems for other reasons.

Involving people increases the opportunities they have to become more knowledgeable about their hazard-scape and the consequences that could occur should a hazard event occur, increases their acceptance of the need for mitigation and warning and their understanding of the limits of their effectiveness, enhances compliance (e.g., with building codes), and increases people's responsiveness to warnings. Warning systems enter the risk management armoury as a result of recognition that mitigation measures may not always protect people (see below), necessitating a need for people to be able to respond. Warnings mediate the relationship between hazards and people, in certain cases, by optimizing opportunities for response.

Warnings serve two general functions. The first is to make people aware of a possible, imminent event (often termed a hazard "watch"). The second is to advise people of a need to act. This second function of warnings highlights the need for the development of effective warnings technologies (e.g., to detect hazard activity and issue information about this activity in a timely way) to be complemented with risk management strategies designed to ensure that people understand warnings (e.g., how much time they have to act—see discussion of Response Times in Chapter 1) and are capable of responding to them in timely, effective, and appropriate ways. The latter typically fall under the heading of readiness of preparedness strategies (see Chapters 3–7).

The development of readiness strategies independently of warnings is also necessitated by the fact that some hazards (e.g., earthquakes) can occur without warning.

This book is primarily concerned with the theoretical and practical issues associated with developing readiness and facilitating the sustained ability of people, businesses and institutions to cope with, adapt to, and recover from natural hazard events. However, because mitigation, warnings, and readiness play complementary roles in risk management, a brief overview of the mitigation and warning strategies that populate risk management planning is provided prior to discussion of readiness.

This chapter also introduces the fact that scientists (who provide information on natural processes and hazards), risk management professionals (who develop and deliver preparedness strategies), and the public (who receive risk information and are expected to act on it) think about hazards, risk, and risk management in different ways. Risk management planning needs to be cognizant of this and accommodate the diverse needs, perspectives, and capabilities that different stakeholders introduce into the risk management context.

Risk management is thus conceptualized here as a process in which multiple stakeholders (e.g., scientists, risk management professionals, community members, etc.) make different but complementary contributions to overall risk management. To accommodate the kinds of interdependencies anticipated by this conceptualization, and ensure that the knowledge and actions of different stakeholders make complementary contributions to the development and implementation of risk management policies and practices, effective risk management is built on a foundation of understanding how stakeholders engage with one another. The importance of doing so is introduced in the next section in the context of adopting strategies such as land use planning, structural mitigation, and building codes to mitigate risk.

MITIGATION

Land Use Planning

Where it is feasible and practical to do so, one of the most effective strategies for reducing risk, enhancing people's capacity to co-exist with hazards, and increasing the likelihood of sustained societal functioning should a disaster occur is planning to avoid development in locations that are susceptible to experiencing hazard activity (Burby, Deyle, Godschalk, & Olshansky, 2000). Burby and colleagues discuss how land-use planning affords opportunities for maximizing the utility of development while simultaneously minimizing risk from hazard events. At a practical level, this includes activities such as planning to avoid developing in areas particularly susceptible to hazardous activity (e.g., flood plains, areas susceptible to liquefaction from seismic activity), land-use zoning restrictions that limit population levels and building densities in susceptible areas, or the relocation of existing at-risk development to safer locations (Gregg & Houghton, 2006).

It is important that national and regional legislation on land use take account of land quality. With regard to the serious liquefaction that occurred in the recent Canterbury earthquakes, which accounted for a large proportion of the damage, engineers had predicted that liquefaction would occur in the eastern suburbs and on the banks of the river Avon, exactly where it did occur (Berrill, 1997; Environment Canterbury [ECan] 2004; Gerstenberger et al., 2011). This shows that zones with even high building standards in terms of building above ground may be inadequate for dealing with liquefaction. It is pertinent that since the Canterbury earthquakes, more subzones have been designated in Canterbury, with some low liquefaction regions deemed to be suited for normal building, others requiring additional measures in their foundations to account for moderate liquefaction risk, and others where further building is no longer permitted due to the risk of major liquefaction. Such plans need to be built into legislation for all regions prone to liquefaction—in New Zealand and elsewhere.

While land use planning is arguably the most effective approach to mitigating risk, it is possible to anticipate circumstances in which other strategies need to be included in comprehensive risk management planning. One such circumstance arises when risk management is applied in environments in which considerable development has already occurred in locations that are susceptible to experiencing hazard events (Thomas & Mitchell, 2001). This can reflect, for example, people wanting to capitalize on the aesthetic qualities of living on coastal fringes or taking advantage of cheaper land in flood prone areas (i.e., to take advantage of the benefits afforded by the very natural processes that can become hazardous). For example, people commonly wish to build close to water (rivers, lakes, and oceans) and yet are reluctant to agree to (pay for) measures such as minimum set-backs that protect development against, for example, coastal erosion, storm surge, and flooding (particularly when such events occur infrequently).

Societies and their inhabitants cannot, once development decisions have been implemented, readily make choices about the nature of their hazardscape (e.g., people cannot decide not to have earthquakes). They can, however, make choices about both their level of susceptibility to experiencing loss from their exposure to hazard activity. Risk management professionals can take information about, for example, the intensity, magnitude, and duration of hazard events and, together with input from professions such as engineering, planning, and architecture, develop strategies to mitigate risk to safeguard people against the particular kinds and levels of hazard activity likely to occur in a given area. Under these circumstances, mitigation can include protecting assets from certain types of hazard activity (structural mitigation) and/or increasing the level of protection offered by buildings (e.g., building codes).

Structural and Engineered Mitigation

Structural mitigation includes engineered measures such as building dams, levees, and sea walls to protect the built environment from flooding, storm surge or, tsunami hazards. Structural measures, such as earth dams, have also been used to protect against lava flows in Hawaii (e.g., Kapoho). Barberi, Brondi, Carapezza, Cavarra, and Murgia (2003) reported the successful construction of barriers on Vesuvius to divert lava flows and afford protection of assets of economic and public importance. Similar activities have been implemented in response to the necessity of dealing with hazard events. For example, in Heimay in Iceland, residents used water cannons to cool and solidify lava flows and prevent them from engulfing the town and destroying the harbor (and thus preventing the loss of an essential and irreplaceable economic resource). However, it should be noted that this strategy has enjoyed mixed success (Fisher, Heiken, & Hulen, 1998).

The level of protection offered by structural mitigation is never total. The design process includes making judgments about levels of protection based on cost benefit analyses that integrates hazard (e.g., return periods, intensity) and political, economic and social considerations (e.g., based on what people would be willing or likely to pay for and maintain). This can result in protection against significant levels of hazard activity, but only up to a point. Consequently, it is possible to anticipate events whose intensity or duration would exceed these design parameters and so, in certain circumstances, expose residents to hazard events. For example, a protective measure may function effectively when experiencing a 100-year event; yet fail catastrophically if a 500-year (typically more intense) event is experienced. Sheaffer, Roland, Davis, Feldman, & Stockdale (1976) discussed how some two-thirds of national losses from flooding in the USA resulted from events that exceed the performance limits of engineered works assumed by most people to ensure their personal safety. To the list of possible problems with structural mitigation can be added issues of construction quality and maintenance that could affect the integrity of the construction and result in levels of protection being lower than anticipated or expected.

Assuming structures are built to specification, the planners and risk management professionals who develop these strategies will recognize these limits (e.g., with this recognition mobilizing the inclusion of readiness planning to complement mitigation initiatives). Members of the public may not, however, apply comparable levels of logic to their analysis of the level of safety proffered by structural measures. It is thus possible to anticipate how people may overestimate the effectiveness of structural measures. This possibility has significant implications for people's beliefs regarding whether they need to do anything to prepare themselves (e.g., see Risk Compensation, Chapter 4).

The ensuing beliefs that people hold regarding the level of protection afforded by a structural measure, particularly when hazard events occur infrequently, can, over time, result in their deciding to live or develop in areas (e.g., flood plains) susceptible to experiencing consequences from hazard events at the higher end of the intensity and duration spectrum. Hence, it is possible for structural mitigation to occasionally have the paradoxical effect of increasing rather than decreasing risk as a result of the people's actions reflecting the influence of human decision processes (such as risk compensation) that tend not to be considered in formal planning analyses and decisions. Consequently, engaging members of the public and community groups in discussions about mitigation is important (see Chapters 6 and 7). It provides a medium for advising people of the limits of effectiveness of structural measures and the need for households and communities to prepare. Another option available to support the implementation of mitigation policy involves the societal establishment of codes and standards for the construction or retrofitting of buildings in areas susceptible to experiencing hazard events.

Building Codes

If societal development has already occurred in area susceptible to experiencing significant levels of hazard activity, steps can be taken to mitigate hazard consequences through the development of design and building characteristics. For example, the difference between the earthquake in Spitak, Armenia, in 1998 where 25,000 people died and the earthquake in Loma Prieta, California, USA, where 57 people died, was not due to the magnitude or type of the earthquake, as the earthquakes had similar magnitudes in both cases. The difference was due to the building design; in Spitak, most of the buildings that collapsed were poorly constructed apartments and houses. In contrast, in California, most buildings had been constructed to withstand earthquakes and more than half of those who died were crushed by the collapse of one bridge that was known to need replacement or strengthening.

Building codes can reduce the adverse impacts of hazards (Gregg & Houghton, 2006). These authors discuss how hurricane clips may prevent roofs from detaching from buildings during high winds, ensure that the roof remains in place to protect from rainfall, and increase the likelihood of a building remaining available for people's use after the event. Similarly, specifying building codes to include, for example, base isolation offers the prospect of a building surviving earthquake shaking.

As with their structural counterparts, however, building codes are effective up to a certain limit of hazard intensity or duration (Godschalk et al., 1999). With any activity introduced to accommodate the implications of events that will occur at some unknown time in the future, it is assumed that, once installed, mitigation measures and devices are maintained to ensure their ability to offer a sustained level of safety over time. This assumption is not always justified. It is important to note that building codes (e.g., standards of retrofitting) may need to be modified as new scientific evidence (e.g., regarding hazard characteristics, engineering techniques) comes to light. Thus, it is possible to anticipate that new information could reveal that measures adopted in the past no longer offer the level of protection or safety believed by planners and owners at the time of building or installation.

If discussion expands to consider how citizens interpret this facet of mitigation, it is worth acknowledging that building codes are designed for a specific function. For example, building codes in seismically active areas are generally intended to prevent buildings from collapsing (and thus increase the likelihood of people surviving an impact) but need not necessarily maintain the structural integrity of the building at a level that ensures its immediate habitability and functionality after a large earthquake event (Gregg & Houghton, 2006). For example, a building may survive an initial earthquake, but fail after experiencing several aftershocks. This was evident in Christchurch, New Zealand, in 2011 following an aftershock that occurred some five months after the initial earthquake. Problems may have been compounded by residents misinterpreting engineering evaluations conducted after the initial earthquake that marked their homes as habitable as indicating its capacity to withstand future seismic activity. In fact, the evaluations referred only to the building's performance to past events and not to the possible effects of future aftershocks on that building. People may overestimate the level of protection offered by having houses built to code and, consequently, believe they need take no further action to prepare themselves (Becker, Paton, Johnston, & Ronan, in press).

The level of sustained functionality afforded by a building is affected by the choices people make regarding the level of risk they believe they are exposed to and what they are willing to invest to mitigate all or part of that risk. People's perceptions of mitigation costs are influenced by several factors. For example, costs may appear relatively low when included in new homes or commercial buildings (e.g., as a proportion of building costs), but may seem disproportionately high if the expense is incurred when retrofitting buildings. When balanced against the uncertainty associated with people's perception of when, if at all, a return on this investment will prevail, the upfront nature of retrofitting costs may constrain actions (see below).

Because it involves expenditure, particularly when any kind of retrofitting is required, the benefits of building codes can also be affected by procedural and distributive justice bias in public policies (e.g., more likely to be adopted in wealthier areas or by owners rather than landlords) that can contribute to inequity in the distribution of risk across sectors of society (Beatley, 1990; Godschalk et al., 1999). Inequity can be compounded by the fact that those in lower socio-economic groups are more likely to live in areas susceptible to loss and be less able to represent their needs to policy makers.

While structural and engineered mitigation and building codes remain important risk management platforms, it is also apposite to bear in mind that such measures are often designed in ways that involve a trade-off between possible levels of hazard activity and the costs of building or adoption of structural measures. This means that people could experience problems from events at the higher end of the spectrum of hazard intensities, magnitudes, and durations that could exceed the parameters of structural measures designed in part on cost-benefit and political criteria. Accommodating the potential for some events to exceed the design parameters of mitigation measures requires including additional risk-management strategies. One concerns monitoring and warning systems.

Warning Systems and Processes

Warning systems are designed to monitor natural processes and detect potential or actual levels of activity capable of being hazardous to people and property. Warnings are intended to provide an early alert to officials and the public in order to facilitate their timely avoidance of consequences or afford people opportunities to mobilize or take protective actions in ways that help them avoid or minimize the experience of adverse hazard effects (Gregg & Houghton, 2006).

The development of warning processes, the messages they disseminate, and the media used to distribute them must accommodate issues associated with motivating the desired public response for events requiring immediate response (e.g., locally-generated tsunamis, tornadoes, flash floods), those that may not occur for several hours or days beyond the issue of a warning (e.g., distant-source tsunami, hurricanes), and those for which there could be a gap of weeks, months, or longer after the initial detection of hazard activity (e.g., some volcanic eruptions). These examples introduce a need for warning process development to be linked to the precursory periods and response times associated with each hazard (see Chapter 1). To these issues, as with other mitigation measures, can be added the need to accommodate economic (cost-benefit) and political criteria in planning warning processes and to understanding how these factors interact to influence the nature of the warning process developed (Johnston et al., 2005). To ensure they realize their potential, it is also important that the development of warning technologies are complemented with activities to ensure the complementary ability of people to respond to them in timely and effective ways..

The need for greater attention to be paid to developing better links between warning processes and people's preparedness and readiness to respond was evident in Gregg, Houghton, Paton, Johnston, and Yanagi's (2007) study of tsunami warnings in Hawaii. In Hawaii, tsunamis have been devastating historically and a public warning system has been tested monthly for decades. Despite the long history of this warning system, Gregg et al. found that levels of public understanding of the meaning of the warning system remained low (e.g., only some 12% of respondents understood the warning process) and had changed little from levels prevailing in the 1960s (Lachman, Tatsuoka, & Bonk, 1961). This work reiterates the need for warnings systems to not just integrate scientific, technological, and mechanical components, but also to accommodate how people relate to and use warning systems. In particular, it is important that public education and outreach programs place considerable emphasis of facilitating the capacity of people and communities to respond in an effective and timely way should a warning be issued (Gregg et al.; Lindell, 1994).

At the same time, it is important to appreciate that not all hazard events afford opportunities to issue warnings (e.g., earthquakes). Some hazards can occur in ways that significantly reduce the utility of any warning (e.g., local source tsunami that may impact coastal communities within tens of minutes from the time they are generated). Ensuring an effective capacity to respond to warnings and developing people's ability to respond to hazard events that offer little or no warning introduces a need to include a readiness component within risk management programs.

Readiness programs and strategies focus on increasing the degree to which people and communities are prepared (e.g., proactively developing household and neighbourhood emergency plans and resources, ability to work with others to confront local problems, capacity for self-reliance, etc.) in ways that increase their ability to cope with, adapt to, and recover from hazard impacts should be unexpected occur (Paton, 2006). While readiness/ preparedness strategies are aimed predominantly at the public, their effective development cannot be pursued without understanding how people relate to the sources of risk information.

READINESS

The inclusion of readiness strategies within a risk management program brings people to center stage in the planning process. This introduces a need for risk management planning to consider not only how to communicate about hazards characterized by uncertainty (see below), but also to consider how to adapt this process to accommodate the diverse stakeholders (who relate to, interpret, and respond to hazardous circumstances in diverse ways) that participate in the risk management process and who need to act in concert to optimize public safety. Several factors make this a challenging task.

Compared with other public safety issues (e.g., road safety), the infrequent nature of hazard (e.g., earthquakes) activity generally precludes people having opportunities to either gain experience of the hazard consequences they could have to contend with or to test the effectiveness of recommended mitigation and readiness activities for themselves. Instead, they have to rely on expert sources (e.g., government, risk management agencies) for information (e.g., via public education, community outreach programs, the media) that covers hazards, their consequences, their mitigation, and recommendations about what can be done to manage their risk. This reliance on others means that the lives of scientists, risk management experts, public officials, and citizens intersect within the risk management context. However, because the members of these groups rarely interact under other circumstances, risk management processes represent the first time their similarities and differences are highlighted in the same context.

With regard to their similarities, the stakeholders who interact within the context of risk management do so because of at least some level of shared concern about hazards and public safety. However, for this shared interest to contribute to their taking shared responsibility for public safety and making complementary contributions to risk management, the differences in how stakeholders engage with the risk management process and with each other needs to be considered and its implications accommodated in intervention planning, design, and delivery. An important issue here is how stakeholders differ with regard to how they think about preparedness.

For example, a need to prepare is self-evident for agencies with a responsibility for risk management and public education and outreach (Ripley, 2006).

They reach this position based on a level of expertise (e.g., training, access to evidence-based practice) that need not be shared by the members of the public with whom they communicate and whose behavior they are attempting to change. Consequently, members of the public do not necessarily share the objective knowledge or beliefs about risk or its mitigation held by their professional counterparts (Fischhoff, Slovic, & Lichtenstein, 1982; Slovic, 1986). The discrepancy in knowledge, beliefs and ways of interpreting hazard and risk information between experts and citizens increases the potential for miscommunication, misunderstanding, or misinterpretation of information (Grothmann & Reusswig, 2006; Mulilis, 1998; Paton, 2003, 2006b; Recchia, 1999; Slovic, 1986). A consequence of this issue is a poor translation of risk information into behavior change and considerable diversity in the nature and levels of people's hazard preparedness (Dow & Cutter, 2000; Grothmann & Reusswig, 2006; Lindell & Perry, 1992, 2004; Lindell & Whitney, 2000; McGee & Russell, 2003; Paton et al., 2005; Thomalla, Downing, Spanger-Siegfried, Han, & Rockström, 2006; Tierney, Lindell, & Perry, 2001).

It is not, however, enough to know that people do not use scientific hazard information in the same way as scientists and emergency planners. It is necessary to understand why this is the case. Armed with understanding why these differences occur, it becomes possible to develop more effective approaches to public education, risk communication, and public outreach. Pursuing the latter is also fundamental to the goal of ensuring that responsibility for risk management is shared amongst those at risk and that they are all playing their part in managing risk. For this to happen, experts and citizens need to be actively engaged in risk management processes in ways that accommodate their interpretive differences and that seek as far as possible to reconcile them in ways that increase the capacity of experts and citizens to play complementary roles in the risk management process (King, 2008; Lindell, Prater, & Perry, 2006; Paton & Wright, 2008). Discussion of how this can be accomplished commences by considering how people's accumulated knowledge and experiences are encapsulated in ways that help them make sense of their world and make decisions about issues that could affect them. This issue is encapsulated in the mental model concept (see also Chapter 6).

MENTAL MODELS

As people accumulate experience and receive information about the environment over time, these inputs are incorporated into people's mental models (Atman, Bostrom, Fischhoff, & Morgan, 1994; Bostrom, Atman, Fischhoff, & Morgan, 1994; Fischhoff et al., 1982; Johnson-Laird, 1983;

Paton et al., 2006; Zaksek & Arvai, 2004). Mental models are abstract mental representations of the objects, events, and relationships that comprise a person's reality. People's mental models encapsulate the meanings they construct over the course of their life. These mental models are used by people to interpret, explain, and then plan for and make decisions about future events (Atman et al., 1994; Bostrom, Fischhoff, & Granger Morgan, 1992; Werner & Scholz, 2002; Zaksek & Arvai, 2004). This section focuses on how mental models influence interpretation of risk and preparedness. Chapter 6 explores how the development of people's mental models is influenced by their social and community relationships.

Mental models play an important role in how people construct, interpret, and respond to risk information (Severtson, Baumann, & Brown, 2006; Zaksek & Arvai, 2004). For regularly occurring events (e.g., those associated with road safety), there is likely to be a reasonable degree of consistency between people's mental models of events and the actual events they are planning for or have to make decisions about. However, when dealing with the uncertainty associated with infrequently-occurring, complex and potentially highly threatening natural hazard events, the gap between people's mental models and the contexts and circumstances they are being asked to consider, plan for, and make decisions about may be substantial (Basili, 2006; Donovan & Blake, 1992; Jones, 1999; Slovic, Finucane, Peters, & MacGregor, 2004; Bechara, Damasio, Tranel, & Damasio, 1997; Donovan & Blake, 1992; Jones, 1999; Kahneman, 2003; Simon, 1955). Yet, despite the fact that hazard and risk information tends to be novel and fall outside the realm of activities that comprise everyday life, people, at least initially, still rely on (pre-existing) mental models developed to accommodate more commonly occurring routine life events and challenges to make sense of hazard and risk management issues. If the goal of reconciling expert and citizen mental models is to be effectively pursued, it is first necessary to understand how they develop and change.

Developing Mental Models

As new information becomes available, whether actively (e.g., from experience) or passively (e.g., from public education), this new information interacts with people's existing model in two ways. The default option is assimilating information or experience. That is, making it fit with existing knowledge structures (Atman et al., 1994; Bostrom et al., 1994; Zaksek & Arvai, 2004). A propensity to assimilate information (i.e., make it fit into existing structures) can make mental models fairly resistant to change. However, models can and do change. Change can be brought about in several ways.

Change occurs when people experience a significant discrepancy between new information or experience and their existing models. For example, the experience of significant loss from a disaster provides people with
tangible experience of the level of discrepancy between their beliefs about
hazard events and what they are actually like. This can act as a trigger for
people to find out more about preparedness. However, since such experience is neither desirable nor feasible for most people, risk communication
needs to devise alternative methods to challenging people's beliefs and eliciting change in their mental models and to do so prior to hazard events
occurring. One approach would be to get people who have experienced a
hazard event to present information to others. For example, using stories
from those affected by the Christchurch earthquake to inform people in
Wellington about what happened and why preparedness would be important. This information would be difficult to dismiss or ignore and when presented by sources that people can identify with, it is more likely to be attended to. This process can be facilitated by encouraging the recipients to discuss
the new information with others (see Chapters 4 to 7 and 10). The presenta-
tion of new experience or information through socially-mediated channels
may result in qualitative shifts in people's mental models and the development of richer ways of constructing, interpreting, and predicting their hazardous reality. Understanding how to change mental models becomes important in the risk management context as it brings together stakeholders with
initially diverse ways of interpreting and responding to risk.

Hazards and Mental Models

All groups involved in risk management, from the scientists who conduct
hazard analyses, the planners and risk managers who use the output of hazard analyses to develop mitigation and public outreach programs, and the
members of the public who are called upon to act on outreach messages and
prepare are influenced by different mental models and thus how events and
relationships are interpreted. While the hazard-scape is common to all groups,
they do not interpret hazard phenomena in the same way.

The hazard analyses (see Chapter 1), from which knowledge of hazard
characteristics and behavior are developed, are relatively objective processes. However, the processes mobilized to decide what to do with that information (e.g., whether mitigation is possible and what level of mitigation and
preparedness is necessary at physical, personal, and societal levels, etc.) result from the operation of mental models that include hazard, hazard management, political, economic, social, and cultural elements. Stakeholders can
be differentiated with regard to the relative weighting they afford each of

these attributes in their deliberating about how to manage risk. For example, scientists are likely to give the highest weighting to hazard data, whereas risk managers may place greater emphasis on political and economic criteria as, for example, they attempt to reconcile hazard data with the pragmatics of budgetary constraints. Differences in relative foci of interest between groups can create considerable scope for differences in interpretation and misinterpretation amongst different stakeholders (e.g., scientists versus citizens). This may not be readily apparent.

Risk management experts and citizens alike use terms such as "hazard," "risk," "mitigation," and "protection" in conversation and discussion and they hear these terms being used in public education and media coverage of hazard issues. However, this cannot be taken to imply that they all interpret these terms in the same way. Rather, the meanings applied to these terms are imbued with interpretations that become increasingly subjective, and influenced by different social influences (see Chapter 6), as the analysis of the risk management processes moves from scientists to citizens.

For all stakeholders, risk management is a process that involves imposing meaning on uncertain and unpredictable events to a point where decisions can be made. However, this is done in different ways by different stakeholders. Scientists translate their findings about hazard characteristics (e.g., ground shaking. lava flows) into probabilistic statements (i.e., statements that reflect some interpretation) using analyses of hazard behaviors (e.g., return periods, intensities). Based on their training, members of planning and risk management professions use scientific data to estimate what could happen in developed areas; construct their mitigation, readiness, and recovery plans accordingly; and communicate the outcomes of their deliberations to citizens both directly (e.g., via public education) and indirectly (e.g., via the media–who add their own interpretation). Members of the public, in turn, interpret the hazard-scape and the information received from scientific and civic sources (both directly and indirectly via media analysis) to determine whether or not they need to do anything and, if they do, to determine what they should do and when they should do it.

Stakeholder Interpretations

Scientists operate at the most objective level of analysis. However, even when working in the same field, scientists may use different theoretical paradigms to inform their work, interpret data, and determine how to present and quantify risk. The process of using hazard data to construct mitigation and readiness strategies and the subsequent extent of their adoption by societies and their citizens is influenced initially by professional interpretation and subsequently by how citizens make sense of hazard data and risk management processes.

When developing their mitigation and readiness strategies, the risk management, planning and policy professionals who receive scientific hazard data can act on them in several ways. This means that risk data are interpreted in ways that reconcile their objective threat (based on the scientific data alone) with the economic and political interests of the society they serve. Plans and actions can be influenced as much by budgetary as by scientific pressures and by political and public views regarding the level of risk a society is willing to bear and the level of mitigation they are willing to support. For example, risk management choices incur opportunity costs. Money directed to risk management becomes unavailable to support, for example, health, education or community projects. Furthermore, opportunity cost debates in the public arena can frame how citizens evaluate risk management options and affect the relative salience of risk management options (e.g., investment in risk management versus in education or health) and reduce levels of support for mitigation expenditure.

In general, the further the point of decision making is from the production of hazard data the greater is the scope for its interpretation. Consequently, the greatest level of interpretive diversity emerges in members of the public. Faced with uncertainty and limited, if any, hazard experiences to guide their behavior, people's choices reflect the culmination of the operation of several psychological, social, and cultural processes that are independent of the hazards per se (see below and Chapters 3–7). In a climate of uncertainty, these interpretive predispositions can function to drive a greater wedge between the risk beliefs of citizens and their professional and scientific counterparts. For example, in the absence of opportunities for stakeholders to engage with each other to confront a common, shared, and challenging hazard event that could focus their attention, stakeholders are more likely to assimilate information in ways that reinforce views that reflect their existing beliefs and goals (see Chapters 5, 6, and 7). What this means is that those responsible for risk communication need to understand how people's interpretive predispositions influence how people make sense of their hazardous circumstances and incorporate within risk communication programs strategies to counter any bias that their interpretive process introduces into people's thinking (see Chapters 4–7).

This issue can be illustrated using examples of how people interpret frequency of hazard occurrence. Scientists and risk managers impose meaning to this facet of hazard behavior by expressing it in terms of probability of occurrence. As outlined above, citizens may not treat these data in the same way as their professional counterparts.

Underestimating the Risk of Low-Frequency Events

Likelihood information (e.g., from return periods—see Chapter 1) plays a significant role in the risk assessment process. Consequently, understanding how hazard frequency information is interpreted has important implications for how such material should be communicated to the members of the public who are being asked to prepare. In general, people take more precautions for high-frequency events, and take fewer precautions for low-frequency events, even if the low-frequency events carry higher potential losses (Slovic, Fischhoff, & Lichtenstein, 1982). It appears that people edit low probabilities as essentially nil. Research on this issue showed that people edited probabilities of 0.000006 and 0.000003 as equal to nil, unless the former probability was explicitly defined as twice that of the latter (Stone, Yates, & Parker, 1994). People's insurance patterns reflect this preference for recognizing higher frequency risks (Slovic, Fischhoff, Lichtenstein, Corrigan, & Combs, 2000).

This bias toward high-frequency events has also been demonstrated in judgments about natural hazards such as earthquakes. McClure and Sibley (2011) asked people to judge the importance of insuring for floods and earthquakes in a given area when the hazards occurred at different frequencies (1 year, 4, 16, and 64 years). The average annual cost of the damage and the annual cost of insurance were held constant across the different hazard frequencies, yet people in the high-frequency conditions (1 and 4 years) were more likely to take out insurance than people in the low-frequency conditions (16 and 64 years). This bias explains why people are likely to lock their cars and homes to reduce the risk of theft, a relatively high-frequency event that has relatively minor consequences, but do not take minimal steps to prepare for floods or earthquakes, which are infrequent events but can have massive consequences. This bias toward higher-frequency events leaves people more exposed to risk from low-frequency events such as earthquakes. This issue becomes even more important given the fact that probabilistic estimates of low-frequency events say nothing about when an event will occur (e.g., a 100-year event could occur tomorrow).

The bias toward high-frequency events occurs partly because people's thinking tends to have a short time frame and focuses more on the more immediate future than on the long-term outlook. With a short-term outlook, the risks from low-frequency events seem small, even if those events are potentially catastrophic when they do occur. A short-term outlook is evident where people do not wear seat belts in countries where this precaution is voluntary, because they (correctly) perceive the probability of having an accident on a single trip as being very low. This bias about low-frequency events can be countered.

The perceived frequency of hazards is affected by people's experience of hazards, and people who have lived longer in a hazard-prone area tend to expect the relevant disasters more (e.g., DeMan, Simpson, & Housley, 1988). This trend is strengthened where people have been personally affected by a disaster (Heller, Alexander, Gatz, Knight, & Rose, 2005; Jackson, 1981). In the case of earthquakes, this experience factor has limited value, due to the low frequency of earthquakes. Other strategies can, however, be effective.

Research on seat belt use has shown that people increase their use of seat belts when safety messages shift the person's time frame and tell people the probability of having an accident over a whole lifetime, rather than the probability of an accident in a single trip (Slovic et al., 1982). This principle can be applied to the risks from low-frequency natural hazards such as earthquakes (Slovic et al., 1982). If people know the risk of experiencing the hazard over a 25-year period, rather than the risk in a single year, they are more likely to recognize the value of being prepared. The fact that people are being asked to prepare for events that could occur at some indeterminate time in the future has other implications for how people interpret their risk.

Perception of Future Events

The tendency to underestimate the risk from low-frequency events like earthquakes is compounded by people's tendency toward time discounting, whereby people discount costs and benefits of future events relative to current events (Frederick, Loewenstein, & O'Donoghue, 2002). More specifically, there is an asymmetry in people's choice between options over time, where people tend to underestimate (discount) future benefits of an action whereas they overestimate future costs. However, this bias is not immutable and is affected by several conditions and different motives.

This bias has been applied mostly to economic choices where people place a premium on benefits nearer in time to these that are more distant. But it also has implications for hazards, in that people may be less motivated to prepare for future hazards because they discount the benefits of mitigating the outcomes of those events. One way to counter this pattern is to provide an immediate incentive for actions that are designed to have long-term benefits. For example, Governments in Japan, Germany, France, and the USA provided scrap benefits, cash for clunkers, etc. to encourage people to trade in older larger less efficient cars for newer more efficient ones. Incentives can influence behavior by changing the cost-benefit ratios.

Being presented with information about low probability events, particularly if presented or interpreted in ways that project the perceived timing of the next hazard event to some point that is well into the future (e.g., a 50-year event) affects the likelihood of action, because people think differently about present and future events. Trope and Liberman (2003) argue that people conceptualize events likely to occur at some distant point in the future (e.g., an earthquake that could occur in 30 or 40 years' time) in abstract terms. In contrast, people interpret events that could occur in the immediate future (e.g., at the start of the wildfire or hurricane/typhoon seasons) in more concrete terms. Trope and Liberman (2003) propose that an important difference between the abstract versus concrete representation of hazard events derives from their differing implications for people's affective responses. As events take on more abstract qualities, they become less likely to trigger the emotional reactions (e.g., anxiety) that motivate people to act to protect themselves (Gifford Gifford, Iglesias, & Caster, 2009; Paton et al., 2005).

Thinking about future abstract events also affects people's cost-benefit judgments. The costs of mitigating actions are incurred immediately. However, any benefit may not accrue until some indeterminate time that may be well into the future. Uncertain and future benefits get discounted. One way of countering this trend involves influencing people's beliefs about the relationship between costs and benefits.

A high cost-benefit imbalance reduces the likelihood of people implementing protective measures. Having made this decision, based on the dominance of costs in their thinking, people then generate subsequent arguments to reinforce this decision (Johnson, Johnson-Pynn, & Pynn, 2007). It may be possible to counter this temporal discounting by encouraging people to first generate arguments about why preparing is beneficial, rather than presenting them with lists of what they should do (Weber et al., 2007). That is, effective strategies could use cognitive dissonance principles to get people to think about and discuss the benefits of preparing before they consider what specific costs are incurred (which occurs when people are given lists of things to do to prepare). The importance of adopting this approach is heightened by the fact that people face competing demands on their time and resources.

The Relative Salience of Present and Future Events

Unless people hold a belief that a hazard event will occur within the immediate future (e.g., a few months to a year), it is unlikely that they will take steps to mitigate their risk, particularly as they may have other, more immediate and more salient pressures in their daily lives. For example, uncertainty may result in the risk associated with natural hazards being perceived as less salient than other challenges (e.g., crime, health care) in people's daily life (Fox & Irwin, 1998; Hill & Thompson, 2006; Kunruether & Pauly, 2004; Powell, Dunwoody, Griffin, & Neuwirth, 2007; Tierney et al.,

2001; Weinstein, 1989). The reduced salience thus attributed to natural haz-ards can result in people developing only rudimentary hazard awareness and knowledge but not invest in understanding how hazard consequences could be mitigated (Breakwell, 2000; Finucane, 2002). This problem can be com-pounded by prior experience (direct and indirect) of hazard events that did not generate the kind of catastrophic consequences they expected.

Estimates of Future Threats

Underestimates of the likelihood of occurrence of natural processes also derive from citizens' experience of prior instances of a natural event such as a hurricane warning or an earthquake that turned out to be small and led to no harm (Guion, Scammon, & Borders, 2007). People may initially follow correct practice and get under a desk or stand in a doorway when they feel a ground tremor, but after many experiences of ground tremors that lead to no harm, they become casual about the risk. In addition, when citizens evac-uate, they experience personal costs, travel expenses, and disruption of their work and personal lives; when their region subsequently is not touched by a hurricane, they can see that, as events turned out, their actions did not affect their outcome. Under this circumstance, people who chose not to evacuate did not face any danger and may have benefited from staying in place. This so-called "cry wolf" (e.g., Atwood & Mdor, 1998) experience can lead peo-ple to have a false sense of security when the next hurricane is approaching. Effective strategies need to counter this effect by emphasizing the probabili-ties of damaging hazard events and the costs that may result from not acting, and also by using regulation. To do so, authorities need to know more about how to communicate about events characterized by risk and uncertainty.

Communicating Risk and Uncertainty

Risk communicators constantly have to grapple with the fact that while they know the events they are communicating about will occur, they cannot say when they will occur or predict the specific hazard behaviors (e.g., inten-sity, duration, distribution, etc.) that will arise when they do occur. This places significant limits on the degree to which people's uncertainty is re-duced when they are exposed to or receive hazard and risk information. While uncertainty triggers the search for greater understanding in scientists and risk experts, it may not do the same for members of the public. In fact, receiving information that includes uncertainty can provide people with jus-tification for inaction (e.g., de Kwaadsteniet, 2007; Hine & Gifford, 1996), and reinforce people's underestimation of their risk (see above discussion of low frequency events). Yet, a certain degree of uncertainty is an inescapable feature of natural hazards. Scientists are left with the problem of how to present the risk honestly while not promoting misguided optimism and justifying inaction. One issue that is important here relates to how the public interpret scientific language.

Hassol (2008) showed that several terms used by scientists to describe risks and hazards have a different meaning to the general public, leading to what might almost be described as cross-cultural misunderstandings. Hassol's analysis focused on climate change but the same principle applies to natural hazards in general. For example, when scientists use the term uncertainty, the public interprets the term as suggesting the scientists do not know if their science is correct or even if an event will occur. Citizens incorrectly interpret the uncertainty as scientists not knowing whether there is a serious risk at all. It would be better to report a range of probabilities and likely severities.

Similarly, when scientists use the term error, the public interprets it as meaning the model is wrong or incorrect, when what scientists are actually referring to is the level of uncertainty related to a measuring device or model. It would be better to communicate about the higher and lower limits of activity suggested by a model (e.g., projecting the likely path of a hurricane and the range of possible paths that it could follow over time). When scientists use the term bias, the public interprets it as unfair or deliberate distortion, and scientists would be better to refer to the degree of offset in observations from an observed or predicted value. When scientists use the term positive trend, as they do in relation to increases in global warming, the public interprets it as a good trend, and scientists would be better to refer to the trend as an upward trend. Thus clarifying the meaning attributed to terminology could improve the communication of the risks associated with each hazard.

A point not emphasized by Hassol (2008) is that scientists' communication of risk to the public may be more persuasive if they stress that their science reflects converging evidence from a number of sources. For example, in relation to climate change, this could include not only changes in gases in the atmosphere but also systematic changes in the timing of the seasons, such as spring occurring earlier and related changes in patterns of vegetation and the ecology, such as different species growing further away from the equator. When describing change and probabilities, it is also important to consider whether to convey information using words or numbers.

Use Numbers or Words?

In the context of communicating risk to the public, is it better to use verbal concepts or numerical probabilities? Budescu, Broomell, and Por (2009) examined this issue in relation to International Panel on Climate Change

(IPCC) reports about probabilities regarding climate change, but the issues apply to other hazards such as earthquakes and "one in a hundred year" floods. All groups were presented with statements about events with probabilities indicated by the terms likely, very likely, etc., as in IPCC reports, and then judged the probabilities of the events in terms of percentages. The control group was not given any indication of how to interpret the verbal IPCC statements, whereas with the experimental groups, following each verbal description was a range of numerical values indicating probabilities in percentiles, in one group with wider bands of probability (e.g., 20%) and in the case with narrow bands (e.g., 5%). Participants in the two groups with both verbal and numerical information gave probability judgments closer to those given by the IPCC than the control group, and those in the group with narrower ranges were closer to IPCC estimates than the broader range groups.

One implication of these findings is that people interpreted statements by the IPCC as implying less extreme probabilities than intended by the authors. On the basis of these findings, Budescu et al. (2009) recommend that statements about hazards should use both verbal terms and numerical values to communicate different levels of uncertainty. They should also adjust the width of the numerical range of probability to match the uncertainty of the particular events in order to reduce misjudgements of risk. Complications can also enter the risk communication process by the frequent need to communicate about several types of hazards.

Misjudgements of the Relative Risk of Different Hazards

People's concern about risks often bears no relationship to the objective probability of their being harmed by those risks (Slovic et al., 1982). There are several factors that account for this pattern. Hazards differ on several key dimensions, including controllability, dread, catastrophic nature, and voluntariness. Slovic et al. (1982) identified three underlying factors in people's perceptions of hazards. The first factor, *dread,* comprises risk features that are: uncontrollable, globally catastrophic, hard to prevent, fatal, inequitable, threaten future generations, produce feelings of dread, hard to reduce, increasing in number, involuntary, and personally threatening. The second factor, *familiarity,* comprises observability, scientific knowledge, immediacy of consequences, personal familiarity, and lack of novelty. The third factor was the number of people exposed. Hazards rated high on the dread factor included nuclear power, nuclear weapons, nerve gas, terrorism, warfare, and crime. However, only nuclear power also scored low on the familiarity factor. Nuclear power was rated a high risk despite the low annual fatalities ascribed to it, which suggests that the combination of high dread and low familiarity is potent in risk perceptions. Although people estimated average annual fatalities due to nuclear power to be low, their estimate of the potential fatalities in a disastrous year was high.

This research shows that the potential for catastrophe is salient in people's risk rankings, and it is pertinent to ask why natural disasters such as earthquakes are not perceived similarly. Slovic et al. (1982) omitted natural hazards in their hazard list, but Brun's (1992) study on similar issues included natural hazards such as avalanches, floods, hurricanes, and forest fires. He found that although "human-made" risks were characterized by the number of fatalities and dread, the risk associated with natural hazards was predicted primarily by novelty and delayed consequences, and only secondarily by dread. People saw the time frame (i.e., frequency) as more salient for natural hazards, and saw the catastrophe dimension as more salient for "man-made" hazards. Because natural disasters are often low on novelty and in frequency, many people do not see them as high-risk hazards and appear to separate when they occur from their consequences in their thinking. Earthquakes were not included on Brun's list, but his finding with other natural disasters is likely to apply to earthquakes.

Another factor leading to people's misjudgements of the relative importance of different risks is the differential exposure of these risks in the news media (Taylor, 1978). News media display vivid events such as road accidents and hurricanes more often than less vivid but relatively common events that lead to high fatalities, such as people falling off ladders or people dying from a stroke. These differences in news exposure shape people's judgments of the relative importance of different risks (see also Chapter 6).

This tendency is explained by the availability heuristic, which proposes that people assess the probability or frequency of an event occurring according to the salience or ease with which instances of the event can be recalled or imagined (Tversky & Kahneman, 1982). Thus people tend to overestimate the frequency of events that are more available, due to greater exposure in the news media or other sources. Applying this heuristic to risks, when people rate the frequency of death from different sources, they tend to overestimate dramatic and sensational events that are more available in their memory and underestimate the causes of death that are unspectacular and common, such as strokes and asthma (Slovic et al., 1982). An analysis of reporting of newspaper reports of incidents leading to death showed that the newspapers overreported rare and spectacular causes of injury and death and underreported common and unspectacular causes. This bias in reporting was significantly correlated with people's perception of the risks, even when the statistical frequency of the events was accounted for. Slovic et al. note that people's availability judgments are also affected by their personal experience. If people have not personally experienced a flood or earthquakes, as is often the case with these low-frequency events, this hazard is less available in their thinking and consequently their risk assessments. For many people, information that is out of sight is out of mind.

These biases in the ranking of different risks are difficult to counter, but one strategy might be to increase the recognition of the risk of hazards such as earthquakes and floods by simulating the features of the risk that generate more concern. For example, although earthquakes are natural hazards, their effects on humans are greatly magnified through their influence on manufactured structures such as buildings and bridges, whose collapse causes most of the negative outcomes in earthquakes. Another strategy would be to strengthen the advice given to citizens on the relative risk from different hazards in a well-designed circulation of information.

A further reason why people do not prepare for a given risk such as earthquakes is that they are aware of many different risks and at the same time are constantly being enticed to expend their finite financial resources on numerous other attractions and necessities rather than activities to manage their risks. Thus strategies designed to get people to take action in relation to a particular risk such as earthquakes need to show why this particular risk is as worthy or more worthy of people's time and resources than the many other risks and attractions that compete for their attention (see discussion of costs and benefits–Chapter 3).

The task of getting people to recognize the risks posed by natural processes is accentuated by the fact that the short- or medium-term risk of such a possibility in the citizens' immediate geographic area cannot be known with certainty. This type of uncertainty affects people's judgment that they need to take action or be prepared (Sprott, Hardesty, & Miyazaki, 1998).

Warnings about an impending disaster such as a hurricane or flood are based on expert judgments of probabilities of the event occurring within a broad geographical area. In this context, people in a specific location often feel uncertain about their own risk and prospects and do not take appropriate action. Thus effective strategies need to counter this effect by pointing to the importance of probabilities of a disaster and the costs of not acting, and by using regulation. This strategy reflects the precautionary principle, which holds that citizens and authorities should take action to counter the potential threat based on the best science available. This principle holds that when scientists suggest there is a major risk, it is prudent to take preparatory action to reduce or mitigate harm, despite the remaining uncertainties. It is a tragic irony that many people often do not adopt this principle but then take action after they have experienced a disaster such as an earthquake in which many of their society have died and much other damage has occurred, and when the risk is no higher than before the disaster (Helweg-Larson, 1999; Siegrist & Gutscher, 2008; see also Chapter 4). A final issue about basic interpretation concerns whether risk information focuses on when events might occur (likelihood) or what people will experience (consequences).

Judgments of Consequences Versus Likelihood

Many public campaigns regarding earthquakes emphasize the likelihood of an earthquake in a particular region. However, judgments of earthquake likelihood appear to bear no relation to preparedness (McClure, Walkey, & Allen, 1999; Mileti & Darlington, 1995). Some research suggests that people's perception that an earthquake is likely in the short- to medium-term future is correlated with earthquake preparedness (Lindell & Perry, 2000; Paton et al., 2005). Findings are consistent in showing that preparedness is higher among citizens who perceive that they are likely to suffer negative consequences from an earthquake if they do not prepare (e.g., Palm, Hodgson, Blanchard, & Lyons, 1990). This suggests that strategies to increase preparedness, rather than focusing solely on the probability of an earthquake, need to emphasize the likely consequences of a major earthquake and encourage citizens to personalize this risk and apply it to themselves. Presenting people with information about consequences provides a context that is more conducive to encouraging people to consider the need for and benefits of taking action to protect themselves from these consequences and increasing their ability to deal with them. That is, what people can do to prepare themselves. The next chapter introduces what being prepared to deal with hazard consequences means. It does so from two perspectives; what people tend to do and what they should do to become comprehensively prepared for natural hazard events.

Chapter 3

HAZARD READINESS AND PREPAREDNESS

INTRODUCTION

The natural processes and the hazards that can emanate from their periodic activity are implicit and permanent facets of the environments people inhabit. They have no choice about the fact of their existence. They can, however, make decisions about how they choose to co-exist with their potentially hazardous circumstances. One such choice takes the form of a class of strategies that collectively fall under the heading of mitigation.

The mitigation (e.g., building codes) strategies discussed in the previous chapter offer people protection from hazard events, but only up to a point. It is, however, possible to anticipate the possibility that hazard events will occur at an intensity or duration that will exceed the level of protection offered by mitigation measures (that reflects decisions regarding the level of risk a society is willing to accept based on such things as cost-benefit analysis). Furthermore, in some cases this mitigation work may not be done, not fully implemented, or not maintained. Thus it is possible to conceive of conditions in which people will experience hazard consequences even if mitigation work is undertaken. Recognition of this possibility introduces a need for another facet of risk management; one that focuses on readiness.

Readiness strategies focus on developing people's ability to (a) anticipate what they may have to contend with (hazard consequences); (b) cope with, adapt to, recover from hazard consequences, and, particularly in areas that can expect to experience hazard events repeatedly; (c) to learn from these experiences. Readiness strategies are intended to place people and communities in a position to be able to *respond* in planned and functional ways to emergent hazard demands rather than having to *react* to them in more *ad hoc* ways. Getting people to a point where they are able to respond, however, is a challenging task. One major reason why it is so can be illustrated with reference to the relationship between knowing one's risk and taking action to manage that risk.

From a risk management perspective, the identification of a significant source of risk motivates action to manage that risk. The same cannot always be said for the citizen recipients of hazard and risk information. While their infrequent occurrence means that people rarely experience natural hazard activity, people can still be aware (e.g., from public education and media coverage of events) of kinds of damage and loss that could accompany a hazard event. It might be expected (and often is) that this awareness would be sufficient to galvanize people into action and to take whatever steps they could to safeguard themselves and to ensure their ability to cope with the loss and disruption that would occur. This assumption is unfounded.

For example, Paton, Smith, and Johnston (2000) reported that while 92% of respondents in a survey of volcanic preparedness acknowledged their risk, only 6% had done anything to prepare themselves. The size of this risk awareness-action gap (92% to 6%) introduces a clear need to understand why people can acknowledge the existence of a threat yet do nothing or little to manage the associated risk. Understanding why this gap exists and how it can be narrowed is the subject of this book.

In order to understand why this gap exists and how it can be narrowed, two things are required. The first is to define what being prepared means. The second concerns being able to explain why people do or do not prepare. This chapter introduces both of these areas. It first introduces the fact that the risk awareness-action gap can arise because people decide not to prepare. It then proceeds to discuss the fact that when people decide to do something, what they do is characterized by considerable diversity with regard to both the nature and extent of their preparedness. This discussion introduces a need to understand how people interpret or impose meaning on the preparedness measures they are being asked to adopt (subsequent chapters examine differences in preparedness by examining psychological and social processes). The chapter then proceeds to discuss what people should be doing if they are to be comprehensively prepared for natural hazards. The chapter opens by pointing out that one contribution to the risk awareness-action gap reflects some people deciding not to prepare in the first place.

TO PREPARE OR NOT PREPARE: THAT IS THE QUESTION

In the course of conducting a longitudinal study of earthquake preparedness behavior, Paton et al., (2005) found that while most people carry out at least some preparations, a minority did not prepare. A similar picture emerged in a subsequent study of earthquake preparedness (Paton & Johnston, 2008). When asking people about what they had done to prepare, Paton and Johnston included a "will not do this" response option. Analysis of these data revealed 22% of respondents stating that they would not take steps to protect their house; 34% would not secure cupboards; 17% would not secure high furniture to walls; 27% would not develop a household and family emergency plan; and 29% would not secure movable objects (e.g., television) in the home. Overall, some 15% of people would not prepare.

This problem may be more acute when dealing with less frequently occurring hazards, such as volcanic eruptions. Paton (2008) asked people who had indicated that they had not adopted any preparedness measure to specify what they intended to do in the future. People were asked to select between "definitely would," "possibly would" or "*would not do in future*" options. This analysis revealed that 63% of respondents stated that they had no intention of checking their level of preparedness for volcanic eruptions. Some 56% stated that they would not seek information about volcanic risk and 49% would not consider finding out what they could do to prepare for volcanic eruptions. Some 58% of respondents stated that they would not take steps to increase their level of preparedness for volcanic hazards and 71% said they would not become involved with local groups to discuss how damage might be mitigated. While the issue of "not preparing" is one that requires more investigation, the above data suggest that some people actively decide not to adopt measures that could offer them protection against hazard effects and/or enhance their ability to deal with the disruption hazard events create.

Recognition of the fact that some people decide not to prepare has implications for the construction of risk communication and public outreach strategies. For example, it means that, when planning public education, it is not enough to know that levels of preparedness in a given area are low. It becomes necessary to find out if levels are low because some people have decided not to prepare, or if levels are low because people need additional guidance to know what to do. If this is the case, then it may be necessary to develop risk communication strategies both to counter reasons for not preparing and to separate ones for those who need additional guidance (see Chapters 3, 7, and 10). However, many people do prepare, but when they do, it is common to find considerable differences between people with regard in what people do and in when they decide to do it.

PREPAREDNESS: WHAT DO PEOPLE DO?

Most people do undertake some preparations. However, irrespective of the hazard under investigation, such activity is marked by considerable diversity with regard to the nature and number of the actions that people perform. This can be illustrated using data from Paton and Johnston's (2008) study of earthquake preparedness in Napier (New Zealand). Napier residents have a high level of risk acceptance for earthquakes and regularly experience low-intensity reminders of the seismically active nature of their environment. But did this translate into actual preparedness? The answer is mixed.

Amongst Napier residents, the most heavily endorsed items were: having tools to effect minor repairs (72%), having a first-aid kit (72%), a torch (70%), and a portable radio (57%). Note that these more heavily endorsed items can be classed as routinely available items with everyday uses (see below). Interestingly, the number of respondents having spare batteries for their torch and radio was slightly lower, being 61% and 48% respectively. The absence of batteries would reduce the utility of having items such as a torch. If people are deprived of access to power, a single set of batteries would not last long. This discrepancy between the presence of an item and its possible utility over the course of an emergency event introduces a need to critically examine how people think about the actions they undertake. For example, it identifies a need to inquire whether people could be identifying themselves as being prepared because they know a particular item (e.g., a torch) is present in their house rather than from their specifically taking steps to consider what they might have to face in the event of a disaster and taking action to ready themselves for hazard event (e.g., having a supply of spare batteries set aside so they can continue to use the torch over the course of an emergency event). This issue is discussed in more detail below.

People also reported other types of preparedness. On average, 60% of responded reported having three days' supply of dehydrated or canned food, and 53% had set aside 2 litres of water per person for three days. Clearly, despite these being relatively simple actions to adopt, not everyone does so. The same can be said for actions directed at enhancing the physical integrity of the home.

In Napier, the number of households reporting having taken steps to ensure the structural integrity of their home, its roof, and its chimney was 57%, 52%, and 58% respectively. It should be noted that these data ought to be qualified by the possibility that they may reflect people's *beliefs* about what they have done rather than what has *actually* been done. For example, the data do not allow differentiating between those whose responses on house integrity reflect the outcome of a professional engineering inspection or their

own (subjective) observation that the structure appears sound. There is anec-
dotal evidence to support this view, with some respondents commenting on
the fact that their response was based on their belief that their house looked
sound (see also Becker et al., in press).

This introduces another interesting aspect of people's preparedness behavior; the possibility that people's beliefs about their level of preparedness
and their actual preparedness may not always coincide (see below).
Differences in preparedness behaviors were also evident in other respects.
For example, some 50% had fastened their hot water cylinder to prevent it
being dislodged by ground shaking. However, only 26% had securely fastened cupboards with latches, only 35% had secured high furniture to the
walls, and only 15% had secured moveable objects (e.g., television) in the
home. Only 28% had developed a household earthquake emergency plan.
Overall, not only do these data indicate that not everyone prepares, but also
that people tend to be selective in what they do.

A similar picture emerged from a study of wildfire readiness (Prior,
2010). Residents in the peri-urban fringe in high wildfire risk areas of Hobart,
Australia were very likely to engage in activities such as keeping lawns short
(84%), having a long hose capable of reaching all parts of the house and the
garden (83%), and keeping gutters cleaned (77%). However, with regard to
the kind of structural measures that protect the home, only 52% of respondents had checked roof coverings, only 21% had screened eaves and events
with metal fly wire, and only 19% had screened wall openings with metal fly
wire. Even fewer, 11%, had installed metal shutters or screens on their windows. Some 40% had a wildfire emergency plan, but of these, only one-third
included family issues in the plan. As with the earthquake data discussed
above, it is evident that people are more likely to implement relatively easy-
to-adopt adjustments that often overlap with routine "everyday" activities
(e.g., people mow lawns and have long hoses for reasons other than just
preparing for wildfire events). Prior's (2010) data suggest that people are less
likely to adopt adjustments that are specifically involved in preparing for
wildfires (e.g., creating a defensible space) and more likely to report having
adopted measures that have a more general utility in everyday life.

From these studies, a trend is evident. Irrespective of the hazard under
consideration, people are more likely to report having adopted items such as
having stored food and water, etc. (often labelled "survival items"—see below) than they are to report having undertaken measures that enhance the
structural integrity of the home or that enhance "home safety" (e.g., securing
furniture). People's preparedness is rarely comprehensive.

The work discussed in this section illustrate the magnitude of the challenge that is encouraging sustained hazard preparedness. This brief intro-

duction to what people do and do not do identifies a clear need to understand why such variability in preparedness exists. The importance of doing so arises not only from a lack of preparedness representing a clear deficit in public safety, but also with regard to its implications for the demands on society should disaster strike.

Why Do People Decide to Do What They Do . . . or Don't Do?

Clearly, the more comprehensively prepared people are, the better able they will be to cope with and adapt to any disruption experienced from hazard activity and the more quickly they will be able to recover. However, those who prepare comprehensively are in a minority. With regard to the data presented above, it may seem good that some 50–60% of respondents to the surveys discussed above had taken steps to ensure the structural integrity of their home. Notwithstanding, this leaves some 40–50% of people whose home is more susceptible to experiencing damage or loss and reduced habitability in the event of a hazard event occurring. By not taking these steps themselves, people are effectively transferring risk from themselves to society (who will need to accommodate such deficits in response and recovery activities).

Based on the above estimate of their being some 40–50% of households unprepared, and extrapolating to the wider population in large urban centers, this could mean that tens, if not hundreds of thousands, of people may need to be relocated and receive societal support as a result of a lack of preparedness rendering them homeless and less self-reliant following a disaster. Furthermore, the relatively lower levels of adoption of structural adjustments such as securing furniture and fittings (with only some 15–30% having done so, or 70–85% of people being unprepared) increases the number of people who face the risk of injury or death from unrestrained household items (e.g., television sets, bookcases) being thrown about by ground shaking following an earthquake (the main hazard in the area from which Paton and Johnston's data were obtained). The discrepancy between the proportion of respondents preparing by enhancing the physical integrity of their house and those taking steps to secure internal furniture and fittings raises the possibility that people think differently about these activities. For example, it is possible to speculate that people assume that if they believe their house is physically secure, they do not need to do anything with their furniture. However, and particularly for events at the higher end of the intensity spectrum (that could exceed the building code specifications or the actions people themselves have performed), this would leave people with no protection from furniture being thrown about by the effects of ground shaking. What these examples illustrate is that becoming prepared is not an all or none process and that people may think differently about different aspects of preparedness (see below).

When inquiring why this state of affairs occurs, it is pertinent to first consider the possibility that these data, since they were obtained in a cross-sectional study, could be tapping into only one point in the process of people progressively developing more comprehensive preparedness over time. This is a possibility, but one that needs some qualification. Some studies (e.g., Becker et al., in press; Paton, Buergelt & Prior, 2008) found that some people believed that irrespective of how little they had prepared (in an objective sense), they were adequately prepared. Thus some people are satisfied with what they have already done and see no reason (if left to their own devices) to do more.

If people have already formed the belief that they are adequately prepared, they are less likely to attend to risk information or contemplate the possibility that they need to do more. People may overestimate their preparedness just because they have done something. It is therefore important to acknowledge that people may believe they have done enough. It is also important to accommodate this possibility in risk communication and community outreach strategies and develop these strategies in ways that can contribute to the progressive development of people's preparedness (see Chapters 4, 5, and 10).

However, just because people have not demonstrated an inclination to do more does not mean that they are disinterested in doing more. One reason for people's apparent reticence to prepare on receipt of information or encouragement to do so can be traced to how people interpret the costs and benefits of preparing for infrequent hazard events.

Weighing Up the Costs and Benefits

With regard to when people should prepare, the objective answer to such a question is now (if they have not already done so!) or at the very least before a hazard event occurs. Preparing well beforehand is essential for hazards such as earthquakes that occur without warning. Even for those hazards for which some warning may be possible, such as wildfire or storm hazards, preparing in advance is necessary because activities such as creating and maintaining a defensible space and taking steps to secure the home from ember attack or installing window shutters and roof clips, take time and all describe activities that need to be taken in advance of the hazard event occurring. However, this is not always reflected in people's thinking.

In a study of wildfire preparedness, Paton, Buergelt, and Prior (2008) found that while some people routinely prepared at the start of each fire season, others stated that they would not prepare until they perceived that a wildfire threat was imminent. That is, only when dangerous weather (e.g., receipt of fire warning, awareness of hot, dry, windy conditions) and forest conditions prevailed, or when fire was perceived as a direct threat to their property (e.g., smoke visible and coming their way). One reason for this was traced to how cost-benefit considerations influenced people's preparedness decisions. Given that the costs of preparing are immediate, but the benefits not evident until some indeterminate time in the future (i.e., when fire occurs), people believe that waiting until a fire event is imminent makes for a better cost-benefit balance (see also Chapters 4, 5, and 6). However, by adopting this approach, people leave themselves insufficient time to take effective action. This same interpretive process can reinforce people's belief that what they have done is sufficient, and, more worryingly, lead to their overestimating what they have done and what they know.

Overestimating the Adequacy of Current Preparedness

Evidence that people can overestimate their preparedness comes from studies that audited people's responses to preparedness questionnaires (Charleson, Cook, & Bowering, 2003; Lopes, 1992). Lopes (2000) found that people would overestimate their preparedness by inferring a level of current preparedness of the basis of their prior levels of preparedness (e.g., because they remembered storing food and batteries in the past but forgot dipping into these supplies to meet routine need and not replenishing them) rather than from a more objective, comprehensive assessment of their current preparedness.

Lopes (2000) first assessed people's preparedness using a questionnaire completed by each household. This was followed by home visits to audit preparedness and check what was actually present. Lopes found discrepancies between people's reports of what they had done and their actual levels of preparedness. This finding that people overreport their preparedness levels in surveys has been found in other hazard preparedness research (Committee on Disaster Research in the Social Sciences, 2006; Charleson et al., 2003; Emdad Haque, 2000; Grothmann & Reusswig, 2006; Prior, 2010; Thomalla & Schmuck, 2004). Observations such as these highlight the benefits of auditing household preparedness. However, the compliance costs of pursuing this option (e.g., the time and costs of monitoring and checking preparedness over time) reduce the feasibility of doing so. Consequently, self-report data on preparedness should be interpreted with some caution.

Estimates of preparedness can be affected by factors such as gender. Grothmann and Reusswig (2006) found that men tended to overreport their level of preparedness. A similar finding was reported by Prior (2010). People may also overestimate their knowledge of preparedness.

In New Zealand, a list of actions to perform should a volcanic eruption occur is provided inside the cover of the Yellow Pages telephone directory. Using this as a basis for assessing people's hazard knowledge in a telephone survey, Ballantyne, Paton, Johnston, Kozuch, and Daly (2000) first asked participants if they believed they could describe the list of protective actions for volcanic eruptions listed in the Yellow Pages. While 92% of respondents acknowledged the existence of their volcanic risk, only 41% of respondents stated that they thought that they could list these actions. Because these data were obtained over the telephone (i.e., people were required to respond immediately and did not have time to check), asking respondents to name these actions provided a good test of the relationship between people's beliefs about their knowledge and their actual knowledge.

When asked to actually name the actions described in the Yellow Pages, only 6% could actually recall these items. This is considerably less than the 92% who knew of their risk and the 41% who believed they knew what these actions were. From this example, it can be inferred that some people estimate their perceived preparedness based on *knowing where* information could be obtained, but often assume that this equates to actual knowledge. The discrepancy between what people thought they knew and their actual knowledge (41% versus 6%) suggests that people can significantly overestimate their actual knowledge.

The possibility that people may conflate recognition (e.g., basing a response on what they see) with actual knowledge (e.g., what they know they have done) has implications for research and assessment. The existence of this discrepancy could affect the accuracy of the estimates of preparedness derived from using questionnaires or other self-report instruments. The magnitude of the discrepancy that could occur can be gauged from telephone surveys that ask people what they know or have done and do not provide cues about what should be done.

For example, Paton (2008) first invited people to describe what they knew about volcanic hazards and what they had done to prepare for them. These responses are listed in the "unprompted" column in Tables 3.1 (knowledge of volcanic hazards) and 3.2 (their statements about what they had done to prepare). Next people were read a list of items and they were asked to state which of these they had adopted. These are listed in the "prompted" column in Tables 3.1 and 3.2. The discrepancy between the unprompted and prompted responses supports the inference that people may overestimate

Table 3.1.
UNPROMPTED AND PROMOTED KNOWLEDGE OF VOLCANIC HAZARDS

Volcanic hazards identified	*Unprompted %*	*Prompted %*
Ashfall	17	18
Gases	8	22
Bombs	10	20
Earthquakes	7	24
Lava	14	14

their preparedness when given cues about what they should have adopted.

With regard to volcanic hazards, recall was poor for all items, but more accurate with regard to the hazard characteristics more usually linked to volcanic eruptions (ash fall and lava flows), but their unassisted knowledge of other hazard characteristics (e.g., gases) was poorer. This can mean that people will be poor at anticipating consequences such as the effects on those with respiratory problems. Overall, poor recall (unprompted scores) and the possibility that prompted scores represent recognition of an item rather than actual knowledge raise issues regarding the need to treat self-report data cautiously if it is being used for research or evaluation.

If people overestimate their knowledge, expertise, or level of preparedness, they will also overestimate their safety, be less attentive to new information (e.g., because they believe it is not relevant for them), less likely to perceive a need for any additional preparation (e.g., because they believe they have already done enough), and be less likely to alter the level of perceived risk they attribute to a hazard. Such beliefs are easily sustained because the infrequent nature of hazard events means that people have few op-

Table 3.2.
UNPROMPTED AND PROMOTED KNOWLEDGE
OF VOLCANIC PREPAREDNESS

Items	*Unprompted %*	*Prompted %*
Fasten cabinet doors	0.5	13
Fasten water heater	0.5	10
Fasten tall furniture	1	12
Fasten heavy objects	1	12

portunities to challenge these beliefs through actual experience (though experience does not always guarantee better preparedness—see Chapter 2).

People interpreting their preparedness in this way illustrates how people can know of the risk posed to them by hazards in their environment but fail to take steps to ensure they are comprehensively prepared. Structured intervention is thus required to facilitate preparedness. The risk communication and community outreach programs used to do so must accommodate the kinds of interpretive processes discussed here and develop strategies to counter these beliefs and encourage comprehensive preparedness (see below). The pursuit of any strategy to increase comprehensive preparedness must also accommodate the fact that people tend to be selective in what they do.

SELECTIVE PREPAREDNESS

The above studies consistently identified a propensity for people to prefer to adopt items such as having a three-day supply of water, torch, portable radio, etc. over more complex measures such as ensuring the physical integrity of the house and developing family and neighborhood response plans. Some reasons why this may be so are more tangible than others.

For example, the capacity of homeowners to make changes to their properties can reflect their greater interest in hazard preparedness to safeguard their investment and the safety of their family (DiPasquale & Glaeser, 1999). Renters, on the other hand, may see themselves as transitory, which affects their risk beliefs and actions, and they necessarily have a decreased capacity to implement any structural changes to the home. Homeowners are more likely to adopt hazard adjustments than renters (Grothmann & Reusswig, 2006; Prior, 2010) and more likely adopt a mix of structural, planning, and survival adjustments. In contrast, renters are more likely to focus their attention on survival items (e.g., storing water), often because their status precludes their performing any structural actions.

Other reasons for people being selective in what they do can also be found. Two of these, the low-cost hypothesis and the possibility that people conflate every day and emergency measures in their preparedness decision making, are discussed here.

Low-Cost Hypothesis

The tendency for people to adopt items such as having water, a torch, and a portable radio illustrates the operation of the so-called low-cost hypothesis (Diekmann & Preisendörfer, 1992). This hypothesis proposes that people are generally more likely to adopt low-cost, relatively ineffective actions over higher-cost but more effective actions. Preparedness items such as storing three days' supply of water, having a torch, portable radio, and alternative cooking source (e.g., a barbecue) are easy to adopt and low cost actions (though their presence does not automatically imply the existence of spare batteries and spare gas cylinders (which may be less likely in the winter months unless the barbecue is singled out for its emergency value)). One reason why items such as those being described in this section are low-cost items is because they are likely to be present in the home because of their more general use in everyday life (e.g., people have a torch in case of power cuts or a barbecue because they eat outside in the summer). That is, they are items that can be present for reasons that are independent of their potential emergency use.

Conflating Everyday and Hazard Adjustment Items

The previous section introduced how some of the most commonly adopted preparedness items share something in common; they all have some everyday utility. Recognition of this led Paton et al. (2005) to argue that people could conflate the outcome of activities associated with meeting normal household needs (e.g., shopping habits or the presence of items such as portable radios that have multiple day-to-day uses) with their level of hazard preparedness. That is, if an item in a list of preparedness measures (e.g., having three days' supply of food) coincides with an "everyday" activity (e.g., having several days' food supplies in the pantry or freezer), the presence of such items may be interpreted by people as indicating an existing level of preparedness. If people do this, it becomes important to ask whether people are fully aware of or have even thought about the emergency uses of these items or the conditions under which they will be required should a hazard event occur.

Thus, it is possible that these items are identified as being present because they were already in the household rather than from their being accumulated as a result of people actually thinking about what they could have to contend with should a hazard event occur and then specifically identifying what they need to do to increase their capacity to deal with the consequences they could experience in a disaster. This possibility of people conflating every day and emergency items was reinforced in a second study that included asking about emergency food provisions in two different ways.

Paton (2008) first asked people if they had three days' supply of tinned food. Some 67% of respondents answered in the affirmative. However, rephrasing a subsequent question (in the same questionnaire) and asking whether people had *changed their shopping habits to gradually increase their emergency food supplies* revealed a very different outcome. The latter question represents a measure that provides clearer insights into whether people are adopting actions for the specific purpose of preparing for the consequences of hazard activity (i.e., buying food specifically for use only in an emergency). For the second measure, only 16% stated that they were building their food supply gradually.

The discrepancy between these scores (67% versus 16%) illustrates how people may conflate certain routine behaviors (e.g., shopping in bulk and storing several days food in a pantry or freezer) with being prepared because they, at that point in time, met the requirements specified in public education materials. However, it does not mean that they have taken steps either to identify why they need to have emergency supplies or to ensure that they have food available in the event of a hazard event that could prevent their acquiring this resource. Thus they may not have considered the fact that a hazard that does not have warning period (e.g., an earthquake) would leave them reliant on what they have on hand and without appreciation that damage to infrastructure may prevent them obtaining supplies (e.g., damage to roads or damaged buildings would prevent securing needed supplies or that prevent supermarkets from being resupplied etc.). Even where warning is possible, such as preceding a hurricane, rapid depletion of supermarket supplies can leave people vulnerable. Furthermore, a need to search for basic necessities during this warning period could also deflect from other activities people should be performing such as preparing the home for storm conditions and/or planning evacuation in the light of an impending threat and at a time when they should be focusing on evaluating changing conditions. This highlights a need not only to identify what preparedness items people list as being present but also to inquire into the kind of thinking (e.g., meeting an everyday need versus being to deal with the privations associated with a disaster) that lies behind the presence of a preparedness measure.

Paton's second question (which controls for the conflation of routine and emergency preparedness thinking) taps into whether people are specifically undertaking behaviors that will enhance their capacity to cope and adapt during the atypical circumstances likely to prevail in a disaster. An affirmative response to the question about increasing the availability of emergency (rather than everyday) food supplies is more likely to have probed the degree to which people have thought about why they need to prepare in order to deal with hazard consequences (rather than a decision about meeting everyday needs). In addition to revealing issues associated with how people interpret preparedness items in their everyday life, this work calls for more attention to be paid to the wording of questions used to assess preparedness.

A case for including items such as a torch or household groceries or having garden tools in assessments of overall levels of preparedness is not being questioned here. However, a problem arises if people assess their preparedness based on knowing of items already present and not on understanding why they are needed in an emergency. Once an item is "ticked off," people may be less likely to think about hazards, their implications for them and their family, and less likely to ponder their need to do anything further. That is, people's risk of experiencing adverse consequences is increased because they become less likely to engage in a more critical appraisal of what they may have to contend with and what they need to do to increase their coping and adaptive capacity. Furthermore, if their assessment is based on already present items, they are more likely to discount or ignore future public information and assume that they are already prepared when they are not. A similar conclusion was reached by Becker et al. (in press). These findings have important implications for researching preparedness.

Research Implications

The possibility that people can conflate routine and emergency items, and assume items that were acquired to meet everyday needs equates to their being prepared for hazard events raises issues about how people think about their responses to preparedness questionnaires (and how they think about preparedness in general). The critical consideration of how people think about preparedness is an issue with significant research implications. The research implications can be traced to the fact that a key goal of preparedness research is to explain differences in observed levels of people's *hazard* preparedness. If the predictor variables in theories of preparedness can account for significant differences in levels of adoption, they can be used to inform the development of risk communication strategies. Methodologically, however, hazard research assumes that the entire content of the dependent variable (preparedness measures adopted) reflects actions undertaken as a result of people deliberating specifically about taking actions to prepare for hazard events. It is being able to explain how people make decisions about hazard preparedness that allows theory to provide the evidence base necessary to inform the development of risk communication programs.

However, if the items that comprise the dependent variable result from different decision processes (e.g., meeting family grocery needs versus ensuring food will be available in the event of an emergency), then the validity of the data being used to test a theory as a predictor of preparedness may be compromised (Paton et al., 2005) and become less effective as a means of informing the development of programs intended to facilitate preparedness.

That is, it is important that research is seeking to explain hazard preparedness rather than shopping habits. Measures that tap specifically into thinking about hazards and their implications would represent a more valid dependent variable and increase the likelihood that the predictor variables used to test preparedness can inform understanding of preparedness decision making. This, in turn, provides a more robust foundation for the development of risk communication programs.

It may, however, be possible for risk communication and public outreach programs to capitalize on people's tendency to favor low-cost items (that overlap with preparedness measures). Research has found that people more readily undertake actions that are useful in relation to multiple risks and benefits, particularly survival actions such as having a torch or radio (Finnis, 2004). This raises the possibility that hazard preparedness can be developed by focusing on everyday, multi-purpose items and using this as a platform for developing more advanced levels of preparedness over time.

Building Preparedness on a Foundation of Easily Adopted Multi-Purpose Items

Multi-purpose items include those with everyday uses (e.g., having a torch or portable radio). They are items that are thus readily available. From the perspective of encouraging hazard preparedness, the key is expanding how people think about them and why they are important in an emergency. One approach to helping people advance their preparedness involves using a strategy suggested by Lindell and Perry (2000) in relation to earthquake preparation. They suggest encouraging households or businesses to start with the least expensive action, and use the momentum built from people's performance of these actions to provide the impetus for their progressing to adopt other, more expensive, actions (e.g., structural changes, engaging an engineer to check the structural integrity of a house) that offer more protection.

This approach to progressively engaging people is likely to be more effective than the current approaches that typically present people with long lists of recommended actions. For example, presenting people with long lists can be overwhelming and, by drawing attention to the more costly actions, reduce the likelihood of citizens developing their preparedness (Paton, McClure, & Buergelt, 2006). A strategy of engaging people in ways that progressively introduce more complex preparedness actions can help counter the misconception that effective preparation is a matter of selecting a few items (see above and Chapter 9).

This strategy can contribute to the important goal of ensuring that people understand that preparedness comprises a series of interdependent functional activities and that the adoption of all is essential if they are to enhance their ability to cope, adapt to the diverse challenges posed at different phases of the disaster impact, response and recovery experience. That is, to ensure that people are comprehensively prepared. To do so, it is first necessary to understand what being comprehensively prepared means and why it is important.

COMPREHENSIVE PREPAREDNESS

Preparedness strategies are designed with the objective of facilitating the ability of communities, their members, businesses, and societal institutions to enhance their safety and ensure their ability to continue to function under the exceptional circumstances of a disaster. People need to be prepared to deal with the full spectrum of response and recovery demands that unfold over time. People need to be able to deal with direct (e.g., earthquakes and aftershocks) and secondary (e.g., loss of lifelines like water and sewerage services) hazard consequences and contend with challenges emanating from the recovery process (e.g., dealing with government agencies, insurance companies, builders, etc.) for weeks, months, or even longer. Furthermore, they may, as was evident in Christchurch following the 2011 earthquake, have to cycle through response and recovery several times as they deal with the implications of aftershock sequences that prolonged their experience of the disaster. The introduction of a temporal dimension into how disaster recovery is conceptualized is important for two reasons. First, it draws attention to the need to be prepared for a prolonged period of time. Second, it introduces a need to consider how the demands and challenges that people will face change over time.

Changes in Demands over Time

People's experience of hazard consequences is not homogenous. It changes over time. If they are to be comprehensively prepared, people need to anticipate the diverse issues and demands they may have to contend with over time and develop the knowledge, competencies, and support resources necessary to help them cope with, adapt to, and recover from disaster. The dynamic nature of the demands and challenges people encounter can be illustrated by distinguishing between the "impact," "response," and "recovery" phases of the disaster experience. While more seamless in reality, sub-dividing the disaster experience in this way makes it easier to appreciate how different types of preparedness activities play different roles at different stages.

of the overall disaster experience. Being comprehensively prepared starts with being able to withstand direct hazard effects.

Direct Hazard Effects

The foundation for comprehensive readiness is structural preparedness. This refers to the preparedness measures or adjustments adopted to increase the survivability of, and level of protection offered by, the home and property during the period of experiencing hazard effects. What this entails changes from hazard to hazard. For example, for earthquakes, it involves such actions as securing the house to its foundations and strapping chimneys to prevent their collapse from ground shaking; whereas for hurricanes, it involves installing window shutters and roof clips. For wildfires, it entails, for example, using fire resistant building materials and creating a defensible space around the home to protect against ember attack. For volcanic hazards such as ash fall, roof design can affect the level of protection offered by the home.

Despite the fact that the specific measures required vary from hazard to hazard, these actions have something in common. They all act to increase the structural integrity of the home or building and enhance the capacity of the home to offer some level of protection (see Chapters 1 and 2) to its inhabitants. If effective, structural measures can increase (though not always—see Chapter 2) the likelihood that people will have a habitable dwelling during the response and recovery period. This can, in turn, reduce demands on response resources (e.g., reduce the number of people requiring medical assistance, the need for temporary accommodation, etc.) and increase people's availability within a neighborhood to offer mutual assistance and social support to others and to be available to contribute to economic recovery (as customers and employees). Having an intact and habitable home creates the foundation for people being able to take responsibility for their recovery and to play active roles in neighborhood and community response and recovery activities. Good structural preparedness provides the foundation for people' ability to deal with the impact of hazard events on them.

Immediate Impact Phase

During the hazard impact phase (e.g., the first three or so days), people may be isolated from all external assistance and have limited, if any, access to normal community processes and societal resources and functions. Public education materials often identify a need for people to be self-reliant (e.g., have enough stored food and water for all those in the home) for a period of time such as three or five days. This recommendation relates to estimates of the time it could take to recover basic levels of services and utilities and thus the period of time over which being prepared would minimize disruption or discomfort as a result of interruption to normal societal services and functions. However, it remains important to impress upon people that this is an estimate and that more severe impacts could require people to function independently of normal societal resources and functions (e.g., having to use stand pipes, temporary toilet facilities, etc.) for considerably longer periods of time.

During the period of initial impact, the effectiveness of people's coping and adaptive efforts will be a function of prevailing levels of individual/household preparedness (e.g., knowledge of hazard impacts and how to deal with them, having the resources available to assist this) and their capacity for self-reliance (Paton, 2006). Typically, public education materials focus on physical resource needs at this time. This is not enough. Recent disasters have increasingly drawn attention to the fact that people's ability to cope is also influenced by their psychological preparedness.

Psychological Preparedness

Analyses of people's experience of disaster are increasingly highlighting the need for more attention to be directed to psychological preparedness. For example, when reporting on their response and recovery experience, people in Christchurch, New Zealand identified the impact of the lack of psychological preparedness on their recovery as being comparable to the lack of physical and social preparedness (Paton, 2012). A need for psychological preparedness was identified as being particularly important with regard to assisting people cope with the impact of repeat aftershocks (identified as significant setbacks to people's physical, social, and psychological recovery) and in relation to adapting to changes in living conditions, social relationships, and livelihoods (e.g., temporary relocation, being without access to water and sewerage services, disruptions to employment, and so on) throughout the several months over which recovery took place. If psychological preparedness is to be incorporated into risk management planning, it is necessary to identify what it is, how it can be applied in the pre-response, and post-event, recovery and rebuilding stages of disaster recovery, and how it relates to other aspects of preparedness. While its value may be more readily apparent when considered in relation to disaster response, psychological preparedness may also be beneficial in pre- and post-disaster contexts.

Psychological Preparedness Throughout the Phases of Disaster

Psychological preparedness can address issues that arise prior to hazard events occurring. Psychological preparedness could be beneficially applied to managing the anxiety that has often been identified as an impediment to people deciding to prepare (Morrisey & Reser, 2003; Paton et al., 2005). In the post-event period, psychological preparedness could help people deal with, for example, socio-legal processes (e.g., litigation, public inquiries) and media coverage that may persist for months or years and result in people having to relive the causes of their stress and trauma. People may benefit from preparation to deal with the blame processes (e.g., self and other blame, counterfactual thinking) that can affect well-being in this context and that can be divisive in community settings. Ascertaining whether psychological preparedness would be beneficial under these circumstances and determining how, when, and by whom it should be provided is an issue for future research. However, the value of psychological preparedness is more readily apparent in relation to the response and recovery phases of the disaster experience.

What Is Psychological Preparedness?

Psychological preparedness is not about eliminating people's vulnerability to adverse emotional and stress reactions. Rather, it is intended to help people understand how and why they react as they do to the stresses and challenges they could experience in a disaster and to assist them to develop the capacities to manage the associated stress over time. Managing stress not only protects well-being, it also enhances motivation and the quality of decision making (e.g., anticipating what could happen, being better able to appraise information, and make decisions in high pressure circumstances). These competencies combine to enhance people's capacity to sustain a level of functioning in the kind of high-demand circumstances likely to prevail in a disaster.

Morrisey and Reser (2003) discuss psychological preparedness as possessing three essential elements. These are: to anticipate the anxiety and concerns that will arise; to identify uncomfortable or distressing thoughts and emotions that may cause further anxiety; and to find ways of managing the responses so that one's coping capacity remains as effective as possible. Procedures such as stress inoculation training and learned resourcefulness represent strategies that could be used to promote psychological preparedness (Meichenbaum, 1996, 2007; Morrisey & Reser, 2003, 2007; Rosenbaum, 1990).

A key goal of psychological preparedness is to develop the degree to which people possess the competencies and capacities (e.g., knowledge; planning/anticipation; recognition; thinking; feeling; decision making; and the management of one's own thoughts, feelings and actions) that influence their ability to predict, comprehend, understand, and manage the emotional correlates of anticipating (e.g., anticipatory anxiety) and facing challenging circumstances. Psychological preparedness can be enhanced by encouraging people to discuss the challenges they anticipate when they meet to talk about preparedness in neighborhood or community groups (see also Chapters 6 and 7). Being better able to anticipate what they may encounter enhances people's ability to predict, respond to, and exercise control over challenging circumstances and to manage and recover from the associated stress.

Psychological preparedness can readily be incorporated into mainstream preparedness planning. Linking psychological and other aspects of preparedness can enhance levels of psychological preparedness by, for example, helping reduce ambiguity and increasing (perceived) control and social support (all factors that contribute to more effective stress management). For example, explaining how and why preparedness measures work and clarifying the relationship between hazard characteristics and preparedness and developing hazard-related family, neighborhood and community relationships could increase (perceived) control and levels of social support, enhancing psychological preparedness in the process.

The investigation of the nature and application of psychological preparedness as well as whether and how it relates to other ways in which preparedness can be facilitated are other areas that will benefit from additional research. Engagement-based approaches to community outreach that build relationships between neighbors and community members can assist both the development of a sense of collective control (e.g., increased collective efficacy–see Chapter 7) and facilitate the development of social support (e.g., informational, tangible, emotional, and belongingness support) which can facilitate people's ability to deal with the physical, social and psychological demands, challenges, and changes likely to be encountered when responding to disaster.

Response Phase

As the impact phase subsides, people are presented with more opportunities to work with neighbors and other community members to confront local demands (e.g., removing rubble, making homes habitable). At this stage, effective preparation is a function of the degree to which efforts have been directed to developing the capacity of neighbors and community members to work collectively to plan and execute recovery and support tasks (e.g., effect local rescue, provide social support, providing assistance to more vulnerable members of a community, being able to use inventories of skills, etc.). Community-based mechanisms can enhance the effectiveness of collaboration with formal agencies (e.g., emergency services, relief), particularly in regard to the effective mobilization and use of volunteers from within the community. Community mechanisms have additional implications for the provision of social support and can facilitate its use in assisting physical and psychological recovery throughout the course of the disaster experience.

Recovery Phase

As the disaster progressively moves through the response phase, and formal intervention strategies and agencies are fully mobilized, effective preparedness will increasingly reflect the quality of interaction (engagement) between people and between communities and response and recovery agencies (e.g., government departments, NGOs). In addition to continuing to use the mechanisms of community self-reliance discussed above, people's ability to adapt at this stage is influenced by the preparations they have made that enable them to deal with temporary or permanent impacts on employment and livelihood. People may also have to deal with a period of evacuation and the prospect of temporary or permanent relocation. For all of these issues, the degree to which people, individually and collectively in families, neighborhoods and community contexts, have anticipated and planned for the demands likely to be encountered contributes to the effectiveness of their response and recovery.

Breaking down the disaster experience into these stages makes it easier to appreciate how the issues people will have to contend with change to some extent with each phase. It was also apparent that the activities that contribute to effective response and recovery can be differentiated along functional grounds. For example, the knowledge and actions required to appreciate why structural changes to a house contribute to preparedness differ from that required to develop collaborative relationships with neighbors to prepare to be able to respond to local hazard consequences. This issue was introduced earlier with regard to the adoption of low-cost measures, structural measures, and so on. The next section discusses how these different types of preparedness activities can constitute specific functional typologies or categories of preparedness.

FUNCTIONAL TYPOLOGIES

Russell, Goltz, and Bourque (1995) developed a typology of preparation measures that resulted in items being functionally categorized as (a) mitigation or structural actions, (b) survival or preparedness actions, and (c) planning actions. Mitigation or structural actions comprise activities that secure the house (e.g., secure house to foundations, secure roof) and its contents (e.g., securing water heaters, tall furniture, mirrors, and installing latches to secure cupboards) to prevent contents from injuring inhabitants (e.g., as a result of the ground shaking accompanying earthquakes or volcanic eruptions). That is, they function to reduce the risk to buildings and their physical contents from the immediate action of hazard activity (e.g., increase the potential of a house to withstand a certain level of ground shaking) and increase the level of protection offered to those in the house from hazard effects.

Russell et al.'s (1995), second category, preparedness or survival actions, describes those activities intended to facilitate people's capacity to cope with and adapt to hazard consequences and to enhance people's self-reliance during periods of disruption. Survival actions include ensuring a supply of water for several days, having a supply of dehydrated or canned food, a radio with spare batteries, a first aid kit, a fire extinguisher, and wrenches to operate utility valves. Finally, planning activities cover, for example, developing a household earthquake emergency plan and attending meetings to learn about hazards and how to deal with their consequences. Note that the latter also introduces a social quality to preparedness activities.

In a later study, Lindell, Arlikatti, and Prater (2009) conducted a factor analysis of hazard adjustment items (e.g., learned how to shut off utilities, have a transistor radio, have four-day supply of canned food, strapped heavy objects, joined an earthquake-related organization, attended meetings about earthquake hazards). They tried four-, three- and two-factored solutions and settled on a two-factor solution as being the most informative. This analysis failed to confirm the three-factor solution described by Russell et al. (1995).

Lindell et al. (2009) labelled these two factors Direct Action (e.g., learned how to shut off utilities, have four-day supply of canned food, strapped heavy objects) and Capacity Building (e.g., joined an earthquake-related organization, attended meetings about earthquake hazards) respectively. Another way of interpreting the content of these factors is as activities that can be performed by individuals (direct action) and those whose performance entails predominantly interaction (e.g., capacity building) with others (i.e., they are social actions).

Lindell et al.'s (2009) decision to opt for a two factor solution was made in the interests of providing the most parsimonious typology. Revisiting the three- and four-factor solutions, however, permits making additional inferences about how people think about hazard preparedness. There were some consistencies across all three solutions. The most reliable was the capacity building (collective activity) factor. It remained intact across all three solutions. Several adjustments loading on the direct action factor (having a first aid kit, having a fire extinguisher, having four days' supply of food, and four gallons of water) retained this association across all solutions. In the three- and four-factor solutions, the items "learned to shut off utilities," "having wrenches to shut off utilities," and "having a transistor radio" formed one factor. In the fourth-factor solution, "strapping heavy objects," "developing an emergency plan," and "installing cabinet latches" formed a factor. Thus, depending on the solution used, different insights into how people interpret the functions of what they are being asked to consider or adopt can be discerned. While a decision to go for a parsimonious solution is sound, there is also some benefit in exploring whether a more comprehensive analysis provides additional insights into how people not only think about preparedness but also how their thinking leads to their categorizing preparedness activities in different ways.

A similar factor analytic study of people's interpretation of preparation measures (Paton & Johnston, 2007) identified eight factors. These were: emergency kit items (e.g., having a torch, a transistor radio, and spare batteries for both, a first aid kit), physical security of the home (e.g., strengthening the house to increase its earthquake resistance by ensuring that walls and roof will not collapse in an earthquake), household emergency planning (e.g., having a household emergency plan, planning where the family should meet after a hazard event), securing household fittings and fixtures (e.g., "fastening tall furniture to walls," "storing heavy items at floor level," "securing cupboard doors with latches"), having a fire extinguisher and knowing how to use it, response resources (e.g., "having tools to make minor repairs," "having a supply of essential medicines," and "having a portable stove/barbecue"), civil defense emergency planning (e.g., knowing whether civil defense agencies had earthquake protection and response plans), and suburb planning (e.g., knowledge of how earthquake protection and response plans would be applied in the area in which they lived). A study of wildfire preparedness (Prior, 2010) confirmed the structural, planning and survival factors proposed by Russel et al. (1995).

Leaving aside for the moment the differences in the categorizations furnished by each study, the work of Russel et al. (1995), Lindell et al. (2009), Paton and Johnston (2008), and Prior (2010) point to the fact that people impose meaning on the activities and that these meanings differentiate preparedness measures along functional lines. This general finding has both theoretical and practical implications.

Theoretical and Practical Implications

From a theoretical perspective, confirmation of the existence of differences in the content of each functional category identifies a need to ask questions about the antecedents of each type of preparedness. A need to ask this question is bolstered by a brief consideration of what is involved in making decisions about each type of preparedness. For example, decisions to adopt certain survival items are relatively simple (e.g., purchasing water containers and filling them). This is an activity that makes low demands on people's skill and time. Decisions about making a family emergency plan, however, involves acquiring knowledge (e.g., about the hazard-scape and its implications for evacuation) and planning skills (e.g., to cover family evacuation when all members are at home, when members are at work or school, and to anticipate family implications of prolonged relocation, etc.). Developing neighborhood or community response plans require certain social skills, a willingness to commit time to working with others, and collective competencies (e.g., involving local people in meetings, defining local needs, giving people a voice, reconciling issues arising from managing community diversity, and representing needs to agencies) that would not be required to check a house or put together a survival kit.

If the development of different preparedness functions (e.g., survival versus community planning) makes different resource (e.g., time to work with others) and skill or competency (e.g., social skills, planning skills) demands on people, it follows that they could differentiated with regard to the antecedents that precede their adoption and maintenance. This implies a need for preparedness research to treat each functional category as a separate dependent variable (as opposed to the common practice of consolidating all preparedness items into a single dependent variable) and examining whether antecedents or predictor variables differed in their relative contributions to explaining variance in levels of each functional category. From a practical perspective, if such differences were found, it would require different risk communication strategies to be developed to facilitate adoption of each functional category. To systematically examine this, however, it will first be necessary to produce a valid and reliable set of functional categories. Drawing on the kinds of activities that could be required to respond to earthquake, wildfire, and storm hazards, examples of the kind of content that could be examined are summarized in Table 3.3.

The contents of Table 3.3 are illustrative only and are not intended to be comprehensive. At present, the available studies offer little in the way of consensus regarding the nature of these functional categories. Studies also differ with respect to the content of each category. This is not surprising given that

Preparing for Disaster

Table 3.3.
EXAMPLES OF INFORMATION, STRUCTURAL, PLANNING,
COMMUNITY AND COMMUNITY-AGENCY FUNCTIONS

Information	Structural	Planning	Survival	Psychological	Community	Community-agency
Information on local hazard-scape	Create defensible space Secure house to foundations	Asses risk to neigh-borhood	Protective clothing	Anticipate anxieties and concerns and develop ways of managing them	Contribute to community hazad plan	Work with agencies to develop comple-mentary roles
Local risk	Check roof coverings Maintain roof	Assess risk to house	Equipment to deal with embers/ spot fires	Identify and develop coping con-sequences	Identify what could happen with others in the community	Develop mitigation and response plans with agencies
Workplace prepared-ness/impli-cations	Screen under floor spaces	Develop a hazard plan with family	Fire extingush-ers	Anticipat-ing and planning	Get actively involved in community meetings	Invite agency representa-tives to teach about equipment
Warnings information	Fix metal shutters to windows Secure chimneys	Monitor environ-ment Know when to activate plan	Survival kit, medicines and documents	Developing social support	Devlop mutual community support strategies	Build relation-ships with businesses

they include different hazards. Consequently, the analysis of functional differences in predictor variables will need to be conducted hazard-by-hazard. However, the studies cited in this section all, albeit to varying extents, support the possibility of developing functional categories for structural, survival, planning, capacity building, and collaborative preparedness. To this list it is necessary to add a psychological preparedness function. The validity of focusing on these categories is supported by their being consistent with the kinds of functions people who experienced disaster describe as making important contributions to their ability to recover (e.g., Paton, 2012; Paton &

Tang, 2009). Once these categories are more clearly defined, it will be possible to conduct research to examine the relative contributions of intra-personal, social, and societal relationship variables to explaining differences in the level of adoption of each category. That is, to predict preparedness. It is identifying predictors that links preparedness research with the practical task of developing the risk communication and community outreach strategies required to facilitate people's preparedness.

PREDICTING PREPAREDNESS

It is evident from the contents of this chapter that, even when living with the same risks, people differ in what they do to prepare. Some people do nothing. For those who do something, they differ substantially in the nature and number of measures adopted. People are selective, but there is a pattern in this selectivity (see above).

Advancing understanding of hazard preparedness, and thus being able to offer the evidence base required to develop robust risk communication and community outreach strategies requires being able to explain the evident variability in both the kinds of preparedness measures people choose to adopt and the level of their adoption (i.e., the number of items in each category—see above regarding discrepancy between securing the house and securing furniture and fittings).

The preceding discussion of the antecedents of the various functional categories (e.g., survival versus planning) drew attention to the fact that they can be differentiated with regard to the underlying competencies, knowledge-bases, and interpretive processes that can be implicated as predictors of their adoption. For example, a person can make a decision to store water (a decision that requires only low-level knowledge) on his or her own. In contrast, developing a comprehensive family emergency plan requires more sophisticated knowledge and social skills. Yet greater levels of knowledge and social and interpersonal skills can be implicated as predictors of comprehensive community emergency plans.

It follows from this broad overview that the antecedent (that can be anticipated to exist) or predictor variables required to elucidate the observed differences in the nature and level of preparedness can be sourced from research on personal, social, and relationship predictors of behavior change. Consequently, the content of the forthcoming chapters is built on the premise that the observed diversity in preparedness reflects differences in how people, individually and collectively, interpret their circumstances and the information available to them from several sources (e.g., civic agencies, neighbors, members of the communities of which they are members) to inform their preparedness decision making and actions. Chapters 4 to 7 discuss the range of factors that have been identified as being able to play a role in explaining differences in levels of preparedness.

This discussion commences in Chapter 4 which explores how people's preparedness actions (or inactions) can reflect the influence of beliefs and dispositional processes (e.g., denial, cognitive biases) that can operate at largely subconscious or automatic levels. Chapter 5 introduces how social cognitive processes influence how people deliberate about the actions they could take. Then Chapters 6 and 7 explore how the nature and qualities of social relationships in a risk management context (and which encompass people's relationships with each other as well as with formal risk management and government agencies) can affect preparedness. As this discussion progresses from the intra-personal to the social, the need for and the benefits of risk communication and outreach strategies adopting community engagement principle become increasingly evident.

Chapter 4

PEOPLE'S BELIEFS AND
HAZARD PREPAREDNESS

INTRODUCTION

The previous chapter closed with identifying a need to explain why, even when they are facing the same potentially hazardous circumstances, people living in a given area differ with regard to the nature and extent of their preparedness. If reasons for this diversity can be identified, this information could be used to inform the development of strategies to facilitate sustained preparedness. This chapter begins this journey by exploring how dispositional processes (e.g., beliefs such as denial, cognitive biases) that can operate at largely subconscious or automatic levels affect how people interpret their hazardous circumstances and make choices about how they will manage their risk. These factors are summarized in Figure 4.1. Discussion of how these might influence preparedness commences with a review of key dispositional characteristics: locus of control beliefs, fatalism, and denial.

DISPOSITIONAL BELIEFS

Locus of control and Fatalism

People who are fatalistic about a hazard believe that there is no point in doing anything about it; they assume that any action they might take will make no difference and will be a waste of time. This fatalistic attitude undermines their taking preventive action. A key personality dimension relating to fatalism is locus of control, which comprises people's beliefs about the sources of control over their lives (Strickland, 1989). People with an *internal* locus of control believe a person's circumstances are a consequence of their own

73

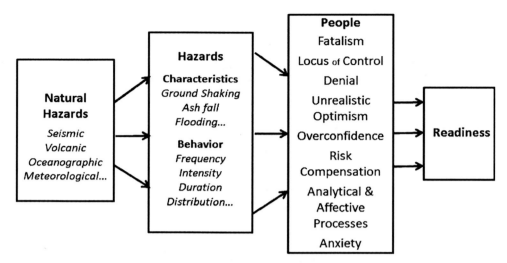

Figure 4.1. Intra-individual influences on preparedness.

actions, whereas people with an *external* locus believe that social forces, such as the government, and chance factors such as fate, determine the circumstances in which people find themselves. These beliefs spill over into people's actions: citizens with an internal locus of control try to exert control over their circumstances more than people with an external orientation.

This pattern applies to preparedness for natural hazards as much as to other parts of people's lives. People with an internal locus of control are more likely to take preparatory action in anticipation of tornadoes (Sims & Baumann, 1972), and to take out flood insurance (Baumann & Sims, 1978). In relation to earthquakes, research has found that an internal locus of control relates to more adaptive perceptions of earthquake hazards (Simpson-Housley & Bradshaw, 1978), and correlates with the view that earthquake damage can be mitigated (McClure et al., 1999).

Given that research has found that an internal locus of control correlates with actions that reduce risk from hazards, clearly it would be useful to change people's locus of control beliefs in an internal direction. This is not a straightforward task, however, because such beliefs are not easily changed and sometimes have strong cultural and psychological roots. Nonetheless, these beliefs can be modified by their being systematically challenged, particularly when interventions are targeted at specific actions, and when they spell out the contingency between mitigating actions and positive outcomes (e.g., Strickland, 1989).

Whereas locus of control involves people's perceptions about their ability to exercise control across all domains of life, more specific measures focus people's views on the controllability of natural hazard events versus the sequences of these events. Hazards such as earthquakes are recognized as ontrollable events, yet their damaging effects can potentially be reduced. ...t direct damage from earthquakes results from building collapse and it has been recognized for some time that it can be explained and reduced by (current) expertise in engineering (Smith, 1993). An important issue in the process of facilitating preparedness therefore is encouraging people to differentiate between natural hazards per se (e.g., earthquakes) and the (more) controllable consequences of hazard activity (e.g., securing houses to foundations and furniture to walls to prevent loss and injury from the ground shaking that accompanies an earthquake). These two categories relate to two different steps in the causal chain of hazard activity and have important implications for understanding how people relate to their environment.

One thing that can happen when people delve deeper into or are exposed to media coverage or information about hazards is that they become more aware of things such as the range of magnitudes they could experience. If, however, awareness of the potential magnitude of hazard threats that could occur interacts with a perceived inability to affect their outcomes should they experience a hazard event, people can respond by developing a sense of numbness or apathy (Gifford, 1976; Macy & Brown, 1998; Moser, 2007; Searles, 1972). Apathy represents an emotional response that can prevent people learning about hazard threat and interfere with their ability to make informed decisions about how they can respond to mitigate that threat.

People's fatalistic perceptions of events as uncontrollable resemble learned helplessness, a passive state that results from people attributing negative events to uncontrollable causes (Abramson, Seligman, & Teasdale, 1978). When people generalize from uncontrollable causes to controllable consequences, they become more helpless. This process may occur with regard to natural hazards.

With earthquakes, for example, this is seen when people assume that because earthquakes are uncontrollable, the devastating effects of earthquakes are also uncontrollable. Turner, Nigg, and Paz (1986) assessed earthquake fatalism in California citizens with statements such as: "There is nothing I can do about earthquakes, so I don't try to prepare for that kind of emergency." Many citizens endorsed these fatalistic items, which express a helpless attitude. People who are fatalistic often fail to differentiate between the uncontrollable nature of the earthquake itself, and the partial controllability of the effects of earthquakes (McClure & Williams, 1996). For example, they fail to see that the quality of building design can mediate the effect of earthquakes (e.g., between shaking and building collapse). Similar principles apply with others hazards, such as flooding.

A positive finding by Turner et al. (1986) was that these fatalistic attitudes did not completely deter people from learning simple strategies that enhance survival in earthquakes. Turner and colleagues found that they could reduce fatalism by getting people to focus on preventing particular instances of damage. They asked citizens if they thought anything could be done to help specific vulnerable groups, such as people living in unsafe buildings, and children in schools. When they focused on these more specific targets, citizens were less fatalistic and believed preventive action could be useful. This finding suggests that one way to overcome hazard fatalism is to focus on specific strategies that mitigate damage and that target specific vulnerable groups. Turner et al. claimed that when they shifted citizens' attention from the fear-inspiring scenario of the earthquake, to helping specific groups using concrete actions, people judged that they could deal with these problems as they were perceived to be of more manageable proportions. These observations are useful in suggesting the most effective strategies for reducing fatalism about earthquakes and for increasing the likelihood of people preparing. They do so by changing the way people make attributions about the causes of damage from hazard activity.

Changing People's Attributions for Damage

The fatalism concept relates to attribution theories, which examine the way different patterns of information shape people's causal attributions for events. The "covariation" theory of attribution proposes that these attributions for events reflect information about the distinctiveness, consensus, and consistency of events (e.g., Kelley, 1967). Applying this model to earthquakes, McClure, Allen, and Walkey (2001) proposed that distinctiveness relates to the damage to buildings in a given earthquake, and consensus and consistency relate to damage in other earthquakes.

Research on this theory has shown how different earthquake scenarios shape people's attributions for earthquake damage. In research on the effect of distinctiveness information, scenarios presenting distinctive damage described a street where a target building collapsed while most other buildings were undamaged. In contrast, scenarios with generalized damage showed the earthquake leading to all buildings in the street collapsing (McClure et al., 2001). As predicted by the theory, people attributed the distinctive damage more to the building's design, a controllable factor, and attributed generalized damage more to the magnitude of the earthquake, an uncontrollable factor.

McClure et al. (2001) examined the effect of consensus and consistency information with scenarios showing whether other earthquakes had led to damage to buildings that were the same design as the target building. The high consensus scenario stated that the target building that collapsed was similar to buildings that collapsed in other earthquakes, and led people to attribute the damage more to the building design. In contrast, the low consensus scenario stated that the building that collapsed was of a type that has usually stood firm in other earthquakes, and led people to attribute damage more to the earthquake.

This result suggests that presenting earthquake damage in the pattern derived from covariation theory can lead people to attribute earthquake damage to building design (in combination with earthquake magnitude). This finding is relevant, because if people attribute earthquake damage to building design, a controllable cause, they are more likely to prepare (McClure et al., 1999). If this is not done, it can increase the likelihood of people denying their risk.

Denial

Another dispositional characteristic that can affect how people relate to the hazardous elements in their environment is denial. When people believe that they have no control over a hazard and/or its activity, this can result in attempts to cope with this aspect of the environment using a mechanism such as denial (Gifford, Iglesias, & Casler, 2008). Denial is a way of coping with an anxiety-producing event where the person denies the seriousness of the risk in order to reduce his or her anxiety. Denial has been shown in relation to natural hazard events. For example, Lehman and Taylor (1988) compared denial in Californian university students who were living in dormitories that were rated as either poor seismically or good seismically. The students were informed of the seismic ratings of their dormitories, and denial of the risk was measured by items such as "Los Angeles was fine in the 1971 earthquake and it will be fine in the next one too." Students in the vulnerable accommodation (who could not influence the characteristics of their dormitories) endorsed these denial items more than students in the sound accommodation.

A related study in New Zealand showed that people's views on the causes of earthquake damage, and the preventability of that damage, were influenced by their degree of exposure to earthquakes and the information they had about the hazard (Crozier, McClure, Vercoe, & Wilson, 2006). Citizens in high- and low-hazard zones either received full information about their zoning (including maps indicating where the soils were likely to shake more) or they received no information about their zoning. In low-hazard zones, citizens who received the zoning information judged that earthquake damage could be prevented more than citizens who received no such information. In contrast, in high-hazard zones, exposure to the same zoning information led citizens to judge that the damage couldn't be prevented. In other words, the knowledge that they were in a high-risk zone led to increased fatalism. This finding suggests that public information schemes in high-risk zones can have counterproductive effects such as leading people to deny their risk or show increased fatalism.

Denial has an inverse relation to the adoption of precautionary measures. For example, research has shown that people's level of earthquake preparedness predicted their estimation of the likely damage from an earthquake (DeMan & Simpson-Housley, 1988). People who had taken fewer precautionary measures underestimated the likely damage to a greater extent, which suggests that people who make fewer precautions cope with the threat of a disaster by denying its likelihood.

How can public agencies counter people's denial of their risk from earthquakes? Individual denial is difficult to change because it serves a functional role in reducing people's anxiety and it is tied into a counterproductive circle with higher exposure to risk. However, agencies can lessen denial. First, they can do so by helping citizens to recognize that earthquake damage can be mitigated. Second, they can assist by both helping people learn that they can have some control, and by increasing their actual control over the hazard and developing their capacity to take relevant action (Lehman & Taylor, 1988; Mulilis & Duval, 1995).

Fatalism and denial are not the only intrapersonal factors capable of systematically influencing how people appraise their hazardous circumstances and make their preparedness choices. Similar influences are exercised by people's cognitive biases.

COGNITIVE BIASES

People's interpretation of hazards and risk can be affected by the subconscious operation of several cognitive biases. These biases can distort people's perception of their relationship with a hazard, affect how they see risk being distributed between themselves and others, and influence the degree to which believe they need to prepare. There are three biases that have more direct relevance for understanding people's preparedness decisions. These are Unrealistic Optimism Bias, Overconfidence, and Risk Homeostasis.

Unrealistic Optimism

Unrealistic optimism (e.g., Weinstein, 1980) refers to a common bias in thinking whereby people think that in comparison with the average person, they are more likely to have a happy future and less likely to suffer misfortunes. This optimism has some benefits such as increasing motivation and persistence; but in relation to hazards, it has significant disadvantages in that it leads people to underestimate their own risk, a bias often referred to as the illusion of personal invulnerability. People know that unfortunate events happen, but they believe that they will not be among those suffering from these events. They think it will happen to someone else.

This optimistic bias has been shown to apply to hazards and disasters. Citizens in Chicago, USA, for example, estimated that if an atomic bomb landed in Chicago, it would kill 97% of the local residents (Burton, Kates, & White, 1993). However, when asked to predict what they themselves would be doing after the bomb exploded, more than 90% believed that they would be helping to bury the dead or taking care of themselves; only 2% thought that they would be dead. In regard to earthquakes, Mileti and Darlington (1995) found that citizens in an earthquake risk zone in the USA expected that an earthquake was likely to occur in the next five years, but they were optimistic that they would not suffer personal loss. Similarly, Mileti and Fitzpatrick (1993) found that although 80 percent of their participants in an earthquake risk zone believed that they would experience a major earthquake, most thought it would not harm them or their property. This effect has been found for different hazards and in different countries.

Sattler et al. (2000) found that most participants believed that an impending hurricane would strike and cause significant damage to their homes, but most believed that the building they lived in could withstand a hurricane. Research has found similar results with citizens in Wellington and Auckland (New Zealand), who judged that they were better prepared for earthquakes and volcanic eruptions respectively, than their acquaintances and others in general (Paton, Smith, & Johnston, 2000; Spittal, McClure, Siegert, & Walkey, 2005). They also thought that they were less likely to suffer harm in an earthquake than were other people they knew.

As Lindell and Perry (2000) noted, these findings show that people who are at risk often fail to personalize this risk. The operation of this cognitive bias effectively results in people transferring risk to others in their community, and this can affect the degree to which they perceive risk information as being relevant to them. For example, while a person may accept that hazard preparedness is important, unrealistic optimism bias can result in their perceiving risk messages that advise of the need to prepare as applying to others but not to themselves (Paton et al., 2000). Thus people may end up inappropriately overestimating their readiness and, as a consequence, reduce their belief in their need to prepare or to attend to risk messages (e.g., because they believe the messages are more relevant for others). Because it provides opportunities to critically assess one's actions against others, this bias is less likely to occur if people are actively engaged in discussing what they are doing to prepare with others (Paton, Buergelt, & Prior, 2008).

A different form of unrealistic optimism can be seen in people thinking that hazards such as earthquakes will happen somewhere else and not in their own city. Recent research examined this pattern following the September 2010 and February 2011 earthquakes in Canterbury, New Zealand (McClure, Wills, Johnston, & Recker, 2011). This study examined participants' recall of their expectancies of an earthquake in their own city and other New Zealand centers before the Canterbury earthquake and their judgments of that likelihood after those earthquakes. The study was carried out in three centers: Christchurch, close to where the earthquakes occurred but where many citizens had not expected an earthquake; Wellington, where an earthquake has been judged more probable; and Palmerston North, which is comparable to Christchurch before the recent earthquakes in that citizens do not (or did not) think an earthquake is likely in their city.

The results showed that in all three cities, expectancies of an earthquake in Christchurch were very low before the earthquakes and very high after the earthquakes, whereas in Wellington, citizens judged an earthquake in their own city as high before and after the earthquakes. In Palmerston North, citizens judged the likelihood of an earthquake low before the earthquakes and moderate (but significantly higher) after the earthquakes. Many citizens in Christchurch said that they had not prepared because they thought an earthquake was going to happen in Wellington and not their own city. This shows that people who live in cities that face lower risk in probabilistic terms may fail to take action to prepare, but their city may be the next one to suffer an earthquake!

Personal Experience

How can this optimism about one's vulnerability be countered? Citizens' perception that they are invulnerable is usually reduced by personal experience of disasters such as floods and tornados that have been personally experienced (Greening & Dollinger, 1992). Burger and Palmer (1992) showed that shortly after experiencing the 1989 earthquake in Loma Prieta, USA, students had no illusions of their invulnerability about being hurt in a natural disaster. However, these students' optimism about disasters in general reemerged three months later. Helweg-Larsen (1999), however, examined optimism about nine negative events and optimism specifically about earthquakes, and found that although people who experienced the 1994 Northridge earthquake showed optimism about the nine other events, they showed no unrealistic optimism about their risk from earthquakes either immediately after the earthquake or five months later. These findings suggest that people who experience a disaster subsequently do not show an unrealistic optimism about their risk from a similar disaster.

Siegrist and Gutscher (2008) replicated this finding in relation to floods, while also demonstrating that people who have experienced a major flood have different perceptions of the consequences of a flood, including the negative emotional consequences. Whereas the non-affected group tended to think that the worst thing that can happen in a flood is destruction of houses and landscapes, the affected group thought that the worst thing is uncertainty and insecurity, a consequence not mentioned by the non-affected group (see also Zakzek & Arvai's 2004 finding in relation to non-experts' understanding of the emotional distress caused by wildfire). Siegrist and Gutscher propose that the reason why affected persons prepared more for a future disaster is that unaffected persons are unable to adequately imagine the emotional distress caused by such a disaster. They also link this difference to the availability heuristic, where aspects of events that come easily to mind (e.g., emotional distress) are given a higher probability and importance. They propose that risk communication should include reference to the likely emotional consequences of natural disasters.

One study found that California adolescents and young adults who had experienced hazards including earthquakes had a *lower* expectancy of dying in a strong earthquake, which conflicts with most other findings on the effects of personal experience (Halpern-Felsher et al., 2001). However, most of the sample had not experienced significant negative outcomes from the hazard. This finding suggests that if people experience a hazard and it has mild consequences for them, they are likely to discount their risk from that hazard in the future. In contrast, citizens who have experienced serious outcomes from a hazard event are less likely to do so. A similar result was reported by Paton et al. (2001) who found that people who experienced low-level volcanic ash fall reported elevated volcanic hazard risk perception but, at the same time, reduced their level of actual preparedness.

Civic bodies that want to encourage citizens to prepare for earthquakes clearly cannot produce sample earthquakes or other hazard experiences to counter this optimistic bias. However, people can be influenced by disasters without being victims of those disasters, particularly if the disasters are salient or relevant (McDaniels, 1988; Taylor & Fiske, 1978).

It is also possible to design outreach strategies to reduce unrealistic optimism. Weinstein (1980) gave people lists of possible precautions taken to reduce particular risks, where the lists had been generated by other people. Compared with a control group who were not shown these lists of precautions, participants' judgments of their chances of experiencing the particular risks showed less unrealistic optimism. These findings show that unrealistic optimism about hazards may be reduced by making people aware of hazards that have harmed other people in similar settings and by telling them about precautions that other citizens have carried out. These strategies may be more effective in changing risk perceptions than taking the apparently rational route of telling people about their risk (Chaiken, 1980).

Overconfidence

Another cognitive bias is seen in people's overconfidence in their judgments, including their predictions of future events (Slovic et al., 1982). People often tend to be overconfident in the accuracy of their risk judgments. Overconfidence can have catastrophic consequences in natural disasters when it affects people's judgments regarding the occurrence of disasters.

This tendency to be overconfident is strongest when accurate judgments are difficult to make. This is just the circumstance people experience when being asked to prepare for infrequent hazard events (Lichtenstein & Fischhoff, 1977). Studies in domains such as medical judgment, financial prediction, and jury decisions show that there is a poor relationship between confidence and accuracy. This bias has not been studied in relation to natural disasters. However, it is clearly important in relation to this domain. Its importance in this regard can be traced to its implying that people often make inaccurate estimates of the likelihood of events such as floods and earthquakes, and that they may hold strong confidence in their predictions even though these judgments are (possibly unknowingly) inaccurate.

How can overconfidence be countered? Experts in domains such as professional weather forecasting, who receive frequent and regular feedback on their predictions, show little or no overconfidence (Plous, 1993). Research applying this finding to non-experts has demonstrated that it is possible to increase the accuracy of people's predictions. This can be done by giving them intensive feedback on the accuracy of their judgments (Plous, 1993).

However, it is not practical to apply this strategy to most real-world situations and particularly to low-frequency events such as earthquakes that do not allow frequent feedback on predictions. It appears that the most effective technique for reducing overconfidence about predictions is to require people to list reasons for and against the predictions they are making (Koriat,

Lichtenstein, & Fischhoff, 1980). This technique has been shown to be effective in countering people's tendency to be overconfident about an option they have chosen, and helps them consider the possibility of alterative futures or outcomes. This strategy could be applied at least to key decision makers in relation to natural hazards, if not the general public.

Another process that can bias people's interpretation arises from a cognitive bias that influences perception of the distribution of responsibility for action. This is known as risk compensation.

Risk Compensation

Risk compensation (Etkin, 1999; Fischhoff, 1995; Lupton, 1999) reflects how people balance their perceptions of how safe the environment is with their need to act to enhance their safety. Risk compensation affects levels of personal preparedness because people erroneously overestimate the effectiveness of mitigation measures (see Chapter 2). For example, the presence of physical mitigation measures (e.g., levees) can lead people to inappropriately exaggerate the level of safety proffered by the environment.

Risk compensation can undermine people's willingness to take responsibility for their own preparedness because it acts to increase the likelihood of people transferring responsibility for their safety from themselves to risk management agencies. For example, Paton et al. (2000) discuss how receiving detailed information about what scientists and civic agencies were doing to manage volcanic risk (through a public education campaign) resulted in some 28% of respondents to a survey on preparedness stating their intention to *reduce* their level of preparedness. This outcome was the opposite of what was intended in giving people the information. Why might this have happened?

One possibility is that in receiving information about what scientists and civic agencies were doing (which they did not have or know beforehand), people came to see the environment as appearing safer than they had hitherto believed (e.g., the risk communication program resulted in people becoming aware for the first time of the depth and breadth of civic risk mitigation activity). Receipt of this information resulted in people suddenly perceiving the environment as safer or as reducing the threat posed by the environment (Paton, Smith, Daly, & Johnston, 2008). This, in turn, stimulated people's belief that they were in a position to reduce their level of preparation (to a level they perceived commensurate with the remaining level of environmental risk).

The net effect of the operation of this bias is a reduction in people's perceived need for personal or household preparedness. This can be particularly problematic in circumstances where societal risk management strategies require (and assume) that people/households are themselves implementing appropriate actions. For example, for wildfire hazards, societal strategies such as controlled burning can go some way to reducing risk. However, risk is further reduced by households developing defensible spaces. However, if people perceive the societal actions as circumventing the need for action on their part, the overall (societal) risk management strategy is compromised.

Strategies for reducing this bias include presenting information in ways that highlight how societal mitigation and personal preparedness complement (rather than being substitutable) one another as part of a comprehensive risk management program, and also highlight that public action is required to cover the fact that societal resources cannot cater for all eventualities (Paton & Wright, 2008). For example, when societal and household actions complement one another, a higher level of risk management will ensue (e.g., the magnitude or duration of hazard activity that can be accommodated by the actions implemented). People's judgments and decisions about hazard issues are also susceptible to being influenced by several decision-making heuristics and by the emotions elicited by hazard issues.

JUDGMENT AND DECISION-MAKING

As with the research on cognitive biases, research on human judgment shows that people's judgments about risks and benefits often deviates from "rational" or "statistical" probabilities (e.g., Tversky & Kahneman, 1982; Slovic et al., 1982). For example, due to the lack of time available for many judgments, or due to an unwillingness to research their judgments, people often base their judgments on short-hand heuristics (rules of thumb) that draw on small but salient samples of information. These heuristics often lead to accurate judgments, but may have been highly adaptive in evolutionary terms and hence it continues to influence people's behavior. However, in many cases, including those relating to judging the risk of a financial investment or the risk from smoking or other hazards, these heuristics and biases can produce inaccurate judgments. To address this issue, research in relation to hazards needs to: first, understand the most common biases in people's judgment in relation to hazards; second, understand which judgmental processes are underpinning these judgments; and third, understand what strategies are most effective for countering these biases as they relate to hazards. Merely supplying people with "rational" information or relevant statistics is often ineffective for this purpose, although it can be part of the solution. First, it is necessary to identify these biases and how they function. This discussion commences with considering how affect or emotion influences decision making.

How Affect Influences Cost-Benefit Judgments

Earlier research on judgment and decision making focused on cognitive factors in heuristics and biases, such as availability (e.g., Tversky & Kahneman, 1982). More recently, researchers have recognized that affect (emotion) plays an important role in people's judgments, including judgments of risk and people's judgments of costs and benefits (e.g., Finucane, Alhakami, Slovic, & Johnson, 2000; Slovic, Finucane, Peters, & Macgregor, 2002). The affect heuristic describes how people's rapid, involuntary emotional response to threatening and challenging events influences their decision making. According to this heuristic, the benefits associated with different events and actions generate an emotion such as feeling good or bad about the event, and this emotion shapes people's judgment of the risks associated with the event. Thus, if people see that an action has a high benefit, they tend to feel good about it and judge the risk of the action to be low. For example, one set of studies on this heuristic showed that when people were told that the benefits of nuclear power are high, they were more likely to infer that the risks were low (Finucane et al., 2000). Changing people's perception of the benefits of an activity changed their judgments of the risk from the activity, even though the objective risk had not changed.

Applied to a natural hazard such as earthquakes, this heuristic implies that when a building has a low price or other positive features, people see the benefits of the low price and feel good about it. As a consequence, they judge the risk of buying the building to be low, regardless of whether the building is unsound in terms of its earthquake resilience. In contrast, if they are aware of the high cost of strengthening the building (see also Chapter 2), this is likely to produce negative emotion, and they judge the financial risks to be high. However, in addition to the reduced risk of earthquake damage that would result from strengthening the building (or buying a stronger one), this heuristic leads them to overlook the benefits of a higher resale value to the building if it was strengthened. This discussion introduces a need to understand how some decisions can result from the application of objective and analytical process while others result from affect-driven processes.

Analytical and Affect-Driven Decision Processes

Theorists have proposed that there are two different systems affecting risk judgments: the more learned cognitive, analytic system, which involves slower computational cognitive processes, and the associative and affect-driven, experiential system, which involves rapid, unconscious affective processes (Chaiken & Trope, 1999; Epstein, 1994; Sloman, 1996). The associative processing system is an evolutionary adaptation that automatically converts uncertain and adverse aspects of environmental experience into affective responses (e.g., fear, dread, anxiety). It thus results in people interpreting risk as an affective or emotional state or feeling (Loewenstein, Hsee, Weber, & Welch, 2001).

Analytic processing, in contrast, is a learnt process that consciously and deliberately applies rules and procedures (e.g., formal logic, utility maximization) to the analysis of data. The analytic system uses algorithms, normative rules and logic, and does not operate automatically. This process is used predominantly by scientists and risk management professionals to quantify their analyses and express the outcomes as probabilities. While ordinary citizens may use some analytical processing, they are more likely to base their decisions on the more readily available associative and affective processing.

This associative and affective experiential system draws on emotions, images, concrete examples, and intuitive associations and often leads to vivid cognitions and emotions. This not only affects ordinary citizens' ability to interpret uncertainty in ways similar to their scientific and professional counterparts, it also increases the likelihood of their experiencing risk in more emotionally-laden ways. Thus if people are exposed to negative emotional labels (e.g., catastrophic) in descriptions of hazards (e.g., through media reporting the use of the "catastrophic" label in Australian bushfire warnings), they are more likely to develop negative beliefs about the hazard.

Even when members of the public do use analytical processes, if the outputs from the two processing systems disagree, the affective system usually prevails (Loewenstein et al., 2001). Analytic processing of scientific information may lead people to recognize hazards as a significant threat, but their affective processing of the output often reduces the likelihood of their acting on this knowledge (Weber, 2006).

Many advocates of logical risk analysis argue that affective judgments about risk are likely to be irrational and harmful, or at least that they lead people to give inappropriate weightings to different risks. However, others argue that experiential affect-based judgments are often adaptive and useful (e.g., Damasio, 1994). For example, the "instinctive" fear that many people feel in a dark passageway or the disgust that they feel to the smell or taste of an unknown substance may protect them from the risks inherent in a given situation. Indeed, many theorists go so far as to argue that the affect heuristic has been the major system of risk assessment enabling survival in human evolution (Finucane et al., 2000). However, there are several limitations to the affective system. One limitation is that affect is easily manipulated by marketing and advertising, often leading to poorer risk judgments. A second limitation is that events that are associated with strong feelings such as terrorist attacks or home invasions can overwhelm people's risk evaluation even when the likelihood of such events occurring is very low. In addition, long-term risks do not trigger affective responses in the way that immediate hazards of threat do, even if those immediate hazards are not significant.

Emotional reactions are critical components of information processing and also have a direct relation to physical and psychological health (Slovic et al., 2004; Groopman, 2004). Certain strong emotional responses, such as fear, despair, or a sense of being overwhelmed or powerless, can inhibit thought and action (Macy & Brown, 1998; Moser, 2007; Nicholson, 2002). Well-intentioned approaches to risk communication based on attempts to create urgency about hazard preparedness by appealing to fear of disasters frequently lead to the exact opposite of the desired response: denial, paralysis, apathy, or actions that can create greater risks than the one being mitigated (Moser & Dilling, 2004). Affect also plays a more active role in decision making when the object of the decision is an unpredictable event believed likely to occur at some point in the future. In this context, anxiety, or its absence, comes to play a role in decision making.

Anxiety

Anxiety is a future-oriented mood state that reflects confronting events that are unpredictable and uncontrollable. Anxiety is accompanied by physiological arousal and by several cognitive reactions, including hypervigilance for threat and danger and elevated levels of fear (Barlow, 2002). Anxiety serves to motivate preparing to cope with future threats, making it a normal, adaptive process unless people are so driven by anxiety that it becomes intense and uncontrollable. In the latter circumstance, the anxiety and worry can become maladaptive (Barlow, 2002; Paton et al., 2005).

There has been relatively little research examining affect and the affect heuristic in relation to natural hazards; however, Hsee and Kunreuther (2000) showed that affect influences decisions whether to purchase insurance. For example, people tend to purchase insurance more for an antique clock to which they are sentimentally attached even if it doesn't work, than to insure against losses when the antique clock does not invoke the same feelings. This finding has obvious extensions to insurance and preparation for natural hazards. The threatening nature of hazard consequences reduces the likelihood of people experiencing this kind of emotional attachment.

Research on affect has also been applied to risk communications about climate change. Many communications that attempt to get people to change their actions to mitigate climate change rely on the analytic system, using graphs or statistics (such as parts per million of carbon in the upper atmosphere). Slovic et al. (2002) proposed that such messages would be more effective if they were complemented by vivid personal experiences and anecdotes about experiences of climate change, such as a river drying up on a farmer's property. However, research suggests that the influence of such appeals on affect is short-lived unless they are reinforced by analytic information (Centre for Research on the Epidemiology of Disasters (CRED), 2009). In addition, frequent exposure to emotional images can lead to emotional numbing, particularly in societies where people are exposed to many images through media such as film and television (Linville & Fischer, 1991).

Furthermore, people have a "finite pool of worry," in the sense that they have limits to their ability to worry about different issues (Linville & Fischer, 1991; Weber, 2006). Thus when there is an economic recession, as in 2008–2009, issues such as natural hazards are more likely to be relegated to a lower level of priority in people's thinking. For example, in a recent study of businesses' preparedness for earthquakes, many companies repeatedly told the researchers that they were merely trying to stay in business, and the risk of an earthquake and the need to prepare for one was at the bottom of their priorities (McClure, Fischer, Charleson, & Spittal, 2009).

People not only have a limited capacity to worry about different issues; when they do respond to a hazard that requires numerous actions, they tend to take only one action and discount the others. This tendency, which is referred to as the "single action bias" (Weber, 1997), is seen where people obtain water supplies to prepare for emergencies and do not secure tall furnishings (see Chapter 3), or when in regard to an environmental issue they recycle material but drive types of cars that degrade the environment. This tendency can be countered by providing checklists that list several of the required actions (but not too many) in relation to a given hazard (CRED, 2009), especially if information is presented in meaningful chunks and presented sequentially from simplest to adopt to the more complex (see Chapter 3).

In a recent project aiming to help businesses to prepare more for earthquakes, the businesses were given en Emergency Management booklet with 64 pages with lists of actions that could be taken. It is telling that many businesses that reported that the booklets were extremely helpful had taken none of the steps recommended in the booklet at follow-up six months later (McClure et al., 2009). Reasons why such discrepancies arise can be traced to understanding how information fits with people's mental models.

People's Models of Natural Hazards

As discussed in Chapter 2, many citizens' mental models of natural hazards and their management differ from expert knowledge and thus contribute to misunderstandings about hazards (Bostrom, Fischhoff, & Morgan, 1992). Communications from experts and civic bodies about hazards to the public often fail to correct these misconceptions. Bostrom et al.'s findings on countering misunderstandings about managing radon risk can be applied to earthquakes and other hazards. This would involve finding out what citizens think happens when a hazard such as an earthquake occurs and then correcting any misunderstanding.

To apply this principle to earthquakes, Hurnen and McClure (1997) examined whether New Zealand citizens' knowledge about mitigating actions related both to their belief that earthquake damage is preventable and to their preparation for earthquakes. Participants' knowledge about mitigating actions was measured by items that experts judged to be useful for preventing earthquake damage (e.g., fastening walls to foundations with anchoring bolts, replacing brick chimneys with metal flues). To control for people giving socially desirable responses, other items described actions that might sound useful to people who knew little about earthquake mitigation, but which would have little real beneficial effect (e.g., build an additional brick wall beside an existing brick wall for added strength). The results showed that participants who were more prepared for earthquakes had higher scores on the earthquake knowledge scale (i.e., they had richer, more sophisticated understanding of cause-effect relationships and the role of adjustments in mediating these relationships–see also Chapter 3).

In this study, each item in the earthquake knowledge scale was explained to participants, spelling out why each action would reduce earthquake damage. After this procedure, participants judged earthquake damage to be more preventable than they did before the study. This finding shows that information about the mechanisms that mediate earthquake effects can enhance people's views that damage can be prevented, and reduce fatalism.

Researchers have also examined experts' and non-expert's mental models about wildfires in a fire-prone forested area in British Columbia, Canada (Zakzek & Arvai, 2004). They organized both groups of participants' mental models into nine domains, including fuel availability and accumulation, ignition sources, conditions for spread, environmental risks, environmental benefits, and quality of life risks. They found that experts held significantly better understanding of wild-land fire in six of the nine areas, including fuel availability and accumulation, ignition sources, conditions for spread, environmental benefits. The two groups did not differ in three areas: fire ignition, risks to the environment, and risks to people's quality of life. In addition, the experts were more experienced in fire prevention than fire management, which the authors see as a deficit, as fire risk can be managed through controlled burn-offs that reduce the likelihood of large uncontrolled fires by reducing stocks of accumulated dry wood. Non-experts also did not understand the reasons for controlled burn-offs, but they had a better understanding of the more human aspects of a fire including emotional distress, which points to an area for improvement in experts' training. In sum, this research reinforces the need to understand what needs to be communicated to people and how it is likely to be interpreted by people before embarking on a specific risk communication program.

These strategies also need to counter the confirmation bias, whereby people tend to seek and recognize information that confirms their existing mental models and to disregard information that is inconsistent with their mental models (see Chapter 2). An example in relation to climate change is where people interpret short-term or local swings in the climate as evidence for their views on climate change, rather than taking account of long-term global trends in climate (CRED, 2009). The confirmation bias also leads people to treat disconfirming evidence as "exceptions to the rule." Another way of countering the action of this (and other) bias involves understanding how to frame the information is presented to people.

The Importance of Framing the Message

When presenting information to people, agencies that disseminate risk information have a choice about how they present information. Information can be presented as abstract facts or lists of items, or information can be framed in ways that can help people personalize it and/or better appreciate its meaning and its implications for them. Indeed, the way that risk communication messages are framed can have significant influences on people's risk judgments and the degree to which the information motivates them to take action.

Research on framing has shown that messages that frame the outcomes of actions in negative terms are more effective than positive frames in leading people to take action. For example, a negatively framed message (e.g., if you do not exercise, you are likely to suffer more illness) has a stronger effect on intentions to exercise than positively framed messages (if you do exercise, you are likely enjoy better health) (Robberson & Rogers, 1988).

Much of this framing research originated in the health promotion literature. However, recent work has extended this framing effect to preparedness for earthquakes. McClure et al. (2009) compared the effect of messages that spelled out the negative effects of not preparing (e.g., if you are poorly pre-pared for a major earthquake, you are more likely to experience harm in the event) with messages that spell out the positive effects of being prepared (e.g., if you are well prepared for a major earthquake, you are less likely to expe-rience harm in the event). Consistent with findings in other domains, the negatively framed messages led to stronger intentions to prepare for earth-quakes than their positively framed counterparts (see also McClure & Sibley, 2011).

One explanation for the greater effectiveness of negative frames is prospect theory, which proposes that people's brains are tuned to be more sensitive to the risk in negative messages (Tversky & Kahneman, 1982). A related explanation drawing on evolutionary psychology argues that nega-tive frames have a greater impact because in past eras, it was adaptive for people to recognize the risks of different courses of action for their survival. These observations suggest that an effective risk communication strategy is to devalue the perceived advantages of hazardous behaviors while promot-ing the benefits of more desirable ones.

Another way to use framing is to focus on people's goals. Some people approach tasks with a "promotion goal orientation" where they are con-cerned with taking actions, advancing toward goals, and maximizing gains; whereas other people have a "prevention orientation" where their main goal is to maintain the status quo, prevent harm from occurring, and minimize losses (Higgins, 2000). Research shows that if messages are tailored to appeal to people's promotion or prevention focus, this increases the probability that people respond positively. Given that civic programs may only have the opportunity to communicate a single message, they need to include appeals to both orientations (promotion and prevention) in the single messages. For example, they could appeal to a promotion focus by emphasizing the bene-fit of an action for increasing the future well-being or prosperity of a com-munity and appeal to a prevention focus by pointing out how the same ac-tions will prevent future harm for a hazard. This strategy requires careful selection of terms that have been shown to represent each goal; for example, terms like *wish, advance,* and *promote,* articulate promotion goals, whereas terms like *minimize losses, avoid mistakes,* and *should,* represent prevention goals. People in the community are more likely to respond positively when messages address their primary goals in promotion or prevention.

Framing can also help people realize that hazards can affect their imme-diate local community, and not only people in other places. For example, Leiserowitz (2007) showed that people in the USA thought that climate change was a serious threat for people in other countries and in other parts of the USA, but not people in their own community. This is another example of the unrealistic optimism effect (where people think that negative events always happen to other people, not themselves) discussed earlier. To counter this bias, risk communication should frame the likely consequences of climate change in terms of local effects in addition to global or national effects. Framing messages can also illustrate predictions about climate change with extreme weather events that are consistent with scientists' predictions about longer-term climate change (CRED, 2009). Framing can also influence the nature of people's cost-benefit beliefs (Finnis, 2004).

This chapter has illustrated how dispositional factors (e.g., fatalism) and cognitive biases (e.g., unrealistic optimism), that operate at an intra-personal level, represent implicit influences on people's interpretation of their hazardous circumstances and their preparedness behavior. It has thus identified interpretive process that have implications for risk communication and which need to be accommodated therein. In the next chapter, the focus turns to examining how social cognitive theories can contribute to accounting differences in people's preparedness behavior.

Chapter 5

PREDICTING HAZARD PREPAREDNESS:
SOCIAL COGNITIVE INFLUENCES

The beliefs and biases discussed in Chapter 4 introduced how factors operating at a subconscious level can influence how people evaluate risk and the options available to manage that risk. This chapter progresses the discussion of factors influencing preparedness by exploring how social cognitive theories can make additional contributions to understanding preparedness. While still focusing predominantly on processes at the individual level of analysis, several of the theories included in this chapter introduce how social context and relationship (e.g., social norms, relationships with information sources) factors can play significant roles in how people make their risk management choices. Before considering the implications of these theories in detail, the chapter first introduces a cognition that plays a pivotal role in understanding the social-cognitive correlates of how people relate to the hazardous elements in their environment: risk perception.

RISK PERCEPTION

For many of the challenges and threats people encounter in everyday life (e.g., road safety), the causes and consequences are generally readily apparent. People may thus have opportunities to gain both direct and vicarious experience of the threats. Regular experience can sustain the risk beliefs that will (hopefully) change people's behavior towards the source of risk in ways that safeguard them from harm. The situation with natural hazards is different; people may not have and may never experience them. This denies people the opportunity to undertake either an objective appraisal of the threat posed by hazard events or the consequences they need to safeguard themselves against. Rather, people have to anticipate the nature and extent of any threat they might experience. Because the phenomena people are required to anticipate rarely afford opportunities for people to experience them directly, the process of assessing threat or risk they pose is open to interpretation (Barnett & Breakwell, 2001; Breakwell, 2000; Caballero, Carrera, Sanchez, Munoz, & Blanco, 2003; Grothmann & Reusswig, 2006; Halpern-Felsher et al., 2001; Jacobson, Monroe, & Marynowski, 2001; Paton et al., 2001a; Proudley, 2008; Weinstein, 1989). As such, it is often referred to as risk perception.

Attributing a level of risk, and recognizing the potential threat a hazard could pose to themselves and what they value is an essential antecedent of the voluntary adoption of protective actions. However, perception of risk itself has not always proven to be a particularly reliable predictor of people taking action to safeguard themselves from harm (Anderson-Berry, 2003; Barnett et al., 2005; Carter-Pokras, Zambrana, Mora, & Aaby, 2007; Davis, Ricci, & Mitchell, 2005; Graffy & Booth, 2008; Grothmann & Reusswig, 2006; Lindell & Whitney, 2000; McCaffrey, 2004a; Miceli, Sotgiu, & Settanni, 2008; Paton et al., 2005; Tierney et al., 2001). Consequently, people's perception of their risk cannot be relied on for motivating behavior change.

For example, in a study of volcanic hazard preparedness (Paton et al., 2000) found that while 92% of respondents acknowledged their volcanic risk, this did not, however, translate into preparedness. Only 6% had developed a volcanic emergency plan, fewer (4%) knew to tape windows to prevent ash entering the house, and less than 1% was aware of the importance of not washing ash into storm drains. While the warning times for volcanic eruptions could allow some time for people to develop their knowledge and implement some actions, these data indicate that knowing of a risk does not necessarily translate into appropriate knowledge or the adoption of actions to manage this risk in the absence of hazard activity (which is when it is most beneficial for people to prepare).

A prominent reason for the risk perception-behavior discontinuity derives from the fact that knowing one's risk does not necessarily equate to knowing how to decrease risk, or how to enact risk reduction strategies. While the relationship between risk and actions to protect from that risk is self-evident to the experts (who know this as a result of their training and experience) charged with communicating with the public, the same cannot be said of the public recipients of this information.

People do not receive special training for making risk management choices under conditions of uncertainty. Rather, to make risk management choices, people rely on the interpretive mechanisms and competencies they have developed and acquired from dealing with the challenges they have

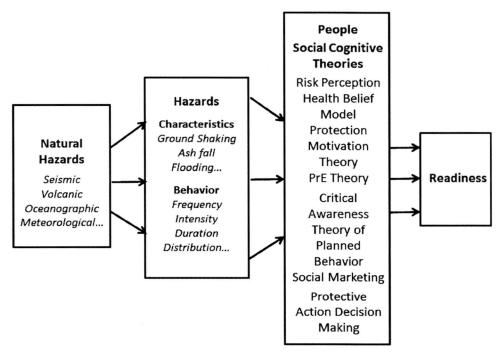

Figure 5.1. Social-cognitive influences on preparedness.

experienced in everyday life. This suggests that one way of enabling action is to identify those mechanisms and competencies and press them into service to facilitate hazard preparedness. This, in fact, is an approach that has been used for several years to facilitate action to manage health risks. How does it work?

People's experiences with dealing with life's challenges and tribulations can be captured in a smaller set of social and psychological variables. For example, the degree of success that people accumulate over time in dealing with day-to-day challenges can be captured by concepts such as self-efficacy and in the coping strategies people develop over time. Recognition of this fact has been used to construct theories that encapsulate how such competencies can inform how people will respond to threats. This work has a proven track record of success in the area of health behavior change (Abraham, Conner, Jones, & O'Connon, 2008). This work defines the starting point for the discussion of how social-cognitive theories can be used to make further inroads into explaining observed differences in people's hazard preparedness. These theories are summarized in Figure 5.1.

THEORIES OF BEHAVIOR CHANGE

Several theories explaining human decision-making and behavior under conditions of risk and uncertainty have been developed, though no one theory is universally accepted. Prominent theories developed in the area of health behavior change are the Health Belief Model (HBM) (Becker, 1974; Rosenstock, 1974); the Theory of Planned Behavior (TPB) (Ajzen & Fishbein, 1988; Ajzen, 1988, 1991); Protection Motivation Theory (PMT) (Rogers, 1975; Bandura, 1977; Maddux & Rogers, 1983); and the Trans-theoretical Model (TTM) (Prochaska & DiClemente, 1982). Several other approaches have built on these theories and applied them more specifically to examining behavior change in response to natural hazard risk (e.g., the Person-relative-to-Event model (PrE) (Mulilis et al., 1990; Mulilis & Lippa, 1990; Mulilis & Duval, 1995), Paton et al.'s (2005) Critical Awareness Model, and Lindell and Perry's (1992; 2000) Protective Action Decision Model (PADM)).

The above theories can all contribute to identifying factors that facilitate behavior change and, importantly, they do so in ways that lend themselves to developing practical interventions (Armitage & Christian, 2003; Kraft, Rise, Sutton, & Røysamb, 2005; Sheeran, Trafimow, Finlay, & Norman, 2002). These theories offer valuable insights into understanding how the discontinuity between recognizing risk and taking action to manage that risk can reflect people not possessing the physical and/or psychological capabilities to search for, adopt, or implement actions to manage their risk (Bennett & Murphy, 1997; Bishop, Paton, Syme, & Nancarrow, 2000; Duval & Mulilis, 1999; Lindell & Whitney, 2000; Paton et al., 2005). They also posit that the risk-action discontinuity can reflect how certain characteristics of people's social relationships can facilitate or inhibit action. For example, people may not act on their risk beliefs if they believe that significant others (e.g., family, friends) would disapprove of the actions they are being asked to perform (Carroll, Cohn, Seesholtz, & Higgins, 2005; Gordon, 2004; McIvor & Paton, 2007; Paton et al., 2008a; Prewitt, Diaz, & Dayal, 2008; Proudley, 2008).

A common denominator between these various formulations is their capacity to provide insights (albeit from different theoretical perspectives) into why people do or do not act in the face of a (known) threat. This confers upon them a capacity to provide a sound evidence-based foundation for developing risk communication, public education, and community outreach strategies that accommodate social-cognitive influences on how people respond to information about events shrouded in uncertainty.

Commonalities between these theories are also evident with regard to their recognizing that simply giving people information about a risk or hazard is insufficient in itself to motivate people to act. Rather, these theories suggest that effective action results from providing people with both high quality information about hazard(s) and their mitigation *and* ensuring they have the personal and social competencies required to interpret and use information to guide their adoption decisions.

Another common denominator between the various social-cognitive conceptualizations of behavior change is their emphasis on the importance of understanding how people weigh the costs of protective action against the benefits that can accrue from their adoption or use. These theories share the goal of operationalizing the social and cognitive antecedents to behavior in ways that contribute to explaining differences in what people do; that is, whether or not people decide to adopt measures or change behaviors in ways that contribute to increasing their safety and/or enhancing their ability to cope with, adapt to, and recover from hazard consequences.

It is pertinent to note that these theories differ in how they propose behavior change; they differ with regard to the number and nature of the variables they include, as well as the inter-relationships between the variables they identify as influencing behavior change. Despite the range of theories that exist, this does not mean that any one theory can be regarded as being more right than any other as such. Rather, they focus on a different part of the puzzle that is hazard preparedness and can, collectively, provide more detailed insights into what a comprehensive set of predictors of hazard preparedness would look like. So, what do these theories have to offer the search for predictors of hazard preparedness?

Identifying a Comprehensive Set of Predictors

The various theories that have been developed offer different perspectives on behavior change. For example, Bandura's (1977; 1988) social cognitive theory suggests that people's acquisition of knowledge and learning is strongly influenced by their observation of the people around them. The TPB describes behavior change in terms of the interaction between attitudes about an action, the subjective norms about the action, and people's perception of the level of behavioral control they believe themselves to have in a given context (Ajzen, 1985; 1991). The HBM introduces into the preparedness environment a need to accommodate people's beliefs regarding the severity of future events, their perception of their susceptibility to experiencing harm from hazard activity, and their opinions regarding the costs and benefits of preparing. Other theories approach the issue of behavior change from different perspectives.

Protection motivation theory posits that fear-arousing communication convinces the receiver that there is a threat, and that this perception instigates "coping appraisal" (adaptive behavior) and "threat appraisal" (maladaptive behavior) with the relative balance between these processes determining whether or not a person acts. The Health Action Process Approach Model describes health behavior change in terms of the development of intentions to act (predicted by risk perception, outcome expectancy, and self-efficacy) and the conversion of intentions into actions, with this conversion being mediated by the possession of knowledge and planning skills essential to ensuring that the desired behavior is performed (Lippke, Wiedermann, Ziegelmann, Reuter, & Schwarzer, 2009; Lippke & Ziegelmann, 2008; Schwarzer, 2008; Wiedemann, Schüz, Sniehotta, Scholz, & Schwarzer, 2009).

While the above-named theories identify the competencies that inform how people make choices, the trans-theoretical (or Stages of Change) model (TTM) argues that behavioral change programs must accommodate differences in people's readiness to act to confront a potential threat. The TTM describes people's level of readiness as a series of stages: pre-contemplation, contemplation, preparation, action, maintenance, and termination that reflect the development of progressively more comprehensive preparedness. An important lesson that can be derived from this theory is the need to tailor intervention to people's level of readiness to change (i.e., one size will not fit all). It also introduces a need to design behavior change programs to engage people at each level of readiness in ways that facilitate their moving progressively from pre-contemplation through to contemplation through to action and to encourage their remaining in the action stage. Armed with this practical knowledge, risk management can then focus on identifying how to tailor strategies to get people who have done nothing to start preparing and ensuring that those who have already prepared maintain their preparedness.

From this brief review, it is evident that, collectively, these theories offer several valuable insights into how people make preparedness decisions and convert these decisions into actions. However, the value of these theories does not rest solely with their identifying the variables that can be used to predict preparedness. Their value can also be discerned in the ways in which they organize the relationships between variables. Articulating the interrelationships between variables serves to reduce the complexity of preparedness behavior by describing complex relationships in more parsimonious ways. In so doing, it provides a framework for developing and delivering risk communication programs that tap into the factors most likely to ensure its success. Thus, to advance understanding of preparedness, it is important to understand what each theory offers the overall understanding of behavior change as it relates to hazard preparedness. Discussion commences with the Health Belief Model.

Health Belief Model

Dooley, Catalano, Mishra, and Serxner (1992) discussed their analysis of earthquake preparedness by relating it to the Health Belief Model (Figure 5.2), specifically with regard to how preparedness reflected the interaction between people's perceived susceptibility to a threat from earthquakes, and their interpretation of the severity of the threat. Dooley et al. confirmed the HBM's prediction that the relative contribution of barriers versus benefits influenced what people did. They also proposed that it would be beneficial to include the TransTheoretical Model (TTM) in the analysis of hazard preparedness, and develop programs that progressively move people from lower to higher levels of preparedness. However, they did not provide any specific insights into how intervention should be tailored to facilitate this progressive transition. Insights into how this can occur can be gleaned from the application of Protection Motivation Theory to behavior change.

Protection Motivation Theory

Protection Motivation Theory argues that behavior change results from the interaction between two processes: one concerned with threat appraisal, and the other with coping appraisal (Figure 5.3). Threat appraisal precedes coping appraisal and involves people accepting that they face a risk and then personalizing it (what it means for them specifically), prior to their considering what they might do to adapt to this risk (see also discussion of PADM below). The strength of a person's threat appraisal is a product of the perceived severity of hazard impacts and people's perceived vulnerability to ex-

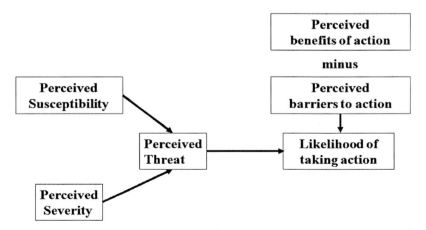

Figure 5.2. The Health Belief Model (adapted from Abraham et al., 2008).

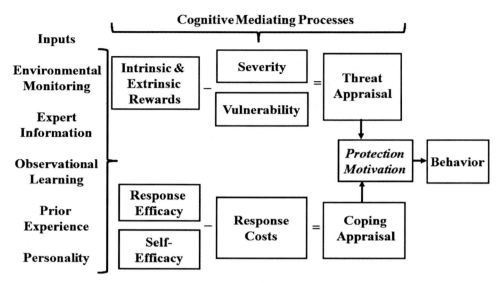

Figure 5.3. Protection Motivation Theory (adapted from Rogers, 1983 and Floyd, Prentice-Dunn, and Rogers, 2000).

periencing adverse hazard consequences. The level of threat appraisal is diminished by the value attached to maladaptive behavior (e.g., financial savings from not acting).

The second process, coping appraisal, is defined by the interaction between response efficacy (i.e., a person's belief about the likely effectiveness of protective measures) and self-efficacy (i.e., people's beliefs regarding their ability to perform protective actions). The latter are weighed against the costs (financial or less tangible entities such as the time, effort, or inconvenience) of engaging in protective behavior. The operation of the sequential processes of threat and coping appraisals culminates in either the adoption or non-adoption of protective measures. Protection Motivation Theory has been used to examine natural hazard preparedness.

Grothman and Reusswig (2006) tested the utility of a model based on Protection Motivation Theory to predict flood preparedness. Their model included threat appraisal and coping appraisal. Coping appraisal comprised three factors: the perceived efficacy of protective responses, perceived self-efficacy (or ability to carry out protective responses), and the perceived costs of protective responses. These perceived costs included time, money, and the effort that would have to be expended to prepare. Protection motivation leads to protective responses which are action-focused and prevent damage in the event of a disaster, with the implementation of protective responses being moderated by barriers such as lack of resources.

Grothmann and Reusswig (2006) concluded that people's perception of their ability to deal with natural hazard consequences is an important influence on their subsequent protective responses. Their findings suggest that information must be provided not only about the risk people face, but also about the costs of measures and how and why they will be effective (i.e., to positively influence benefit-cost ratios—see also Chapter 4). They also argue that changes to the levels of societal support for mitigation (e.g., providing incentives for flood-proofing properties) need to be considered in order to reduce barriers to self-protective behavior and encourage support for and adoption of mitigation measures (see also Chapters 3 and 4).

The PMT has also been applied in conjunction with the Transtheoretical (or stages of change) Model (TTM) to accommodate the fact that, at any point in time, people can be at different stages of readiness to act. For example, at any one time in a given location, some people within the target population may be doing nothing, while others may be considering change. Yet others may have adopted some adjustments, and some people will be performing and maintaining a more comprehensive set of protective actions, and others still may have prepared in the past but have now ceased to do so. Because the factors that motivate people to prepare differ as they progress through the various stages, the application of the TTM proposes that public education and outreach strategies should be designed in ways that engage people at each level and that can facilitate their moving from pre-contemplation, to contemplation, to action, while maintaining levels of preparedness.

The benefits that can accrue from this kind of theoretical integration are evident in studies of flood and wildfire hazard preparation (Bočkarjova, van der Veen, & Geurts, 2009; Martin, Bender, & Raish, 2007a). The combined PMT-TTM model used in these studies includes risk information and risk perception dimensions on the one hand, and stage of readiness on the other, to predict people's motivation to develop hazard preparedness in a progressive manner. These studies reinforce the importance of variables such as perceived risk severity, vulnerability to adverse hazard consequences, self-efficacy, and response efficacy (outcome expectancy) play as predictors of preparedness. However, they also highlight how the relative salience of predictor variables can change depending on people's level of readiness to change.

Determinants of action differ across each stage. For example, for those at the *pre-contemplation stage,* two factors, perceived vulnerability to flooding and response efficacy (also known as outcome expectancy), were the most effective predictors. For those in the *contemplation stage,* response efficacy was the best predictor, with information about the severity of consequences, the costs of protective actions, and subjective knowledge playing marginal roles at this level. Subjective knowledge was positively implicated as a predictor at this level of readiness. Thus, strategies focusing on providing information about hazards, coupled with clear explanation of the effectiveness of proposed measures, could positively influence the likelihood of people in this stage advancing to the action stage.

For those at the *"action" stage,* more attention should be directed to enhancing people's perception of the intrinsic and extrinsic benefits (to both the individual and the society) of engaging in protective behavior. At the same time, less attention should be given to the costs associated with taking protective action (such as time, effort, inconvenience, and money) (Bočkarjova et al., 2009). Finally, when actively engaged in preparing, people's choices are driven by coping appraisal. Furthermore, high levels of response efficacy (positive outcome expectancy) and self-efficacy help sustain preparedness over time (Bočkarjova et al., 2009). These studies demonstrate the utility of Protection Motivation Theory as a theory capable of predicting preparedness. A variant of protection motivation theory developed specifically for investigating hazard preparedness is Person relative to Event Theory.

Person Relative to Event (PrE) Theory

Person-relative-to-Event (PrE) theory (Duval & Mulilis, 1995; Mulilis & Lippa, 1990) predicts the emergence of protective action under conditions of increased fear in the presence of insufficient resources relative to the magnitude of the threat people perceive as emanating from a hazard (Figure 5.4). This is similar to the vulnerability, severity, and response-efficacy elements of PMT, but PrE theory builds on the latter to elucidate the conditions that

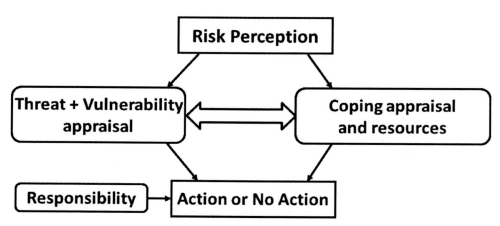

Figure 5.4. The Person-relative-to-Event (PrE) Theory (adapted from Duval & Mulilis, 1995; Mulilis & Lippa, 1990).

promote problem-focused coping in the context of negative threat appeals (i.e., fear-arousing persuasive communication). The PrE model of hazard preparedness is based on the relationship between the appraised level of person resources and the appraised seriousness of the threat, as these are pivotal to understanding the degree to which a person adopts a problem-focused coping approach to confronting hazard events (e.g., by preparing).

The PrE model builds on the assumption that the degree to which a person perceives him or herself at risk from a potentially harmful hazard event interacts with person (self-efficacy, outcome efficacy) and event (severity of event, and probability of occurrence of the event) variables to predict the adoption of hazard adjustments. It thus posits a need to examine person-environment interactions if an understanding of differences in people's preparedness levels is to be understood. The PrE model argues that someone who appraises their resources as sufficient in both quality and quantity *relative* to the magnitude of a particular threat (see Chapter 1) will engage in more problem-focused coping activities than one who appraises his or her personal resources as insufficient relative to the seriousness of the threat (e.g., information on hazard intensity, magnitude, duration, etc.).

The PrE model argues that problem-focused coping influences behavior if the magnitude of a threat is exceeded by the person's resources to deal with the threat. If people believe that their resources are high, then when the threat increases preparedness increases. If people believe that their resources are low, then as the threat increases, their preparedness decreases. The PrE model also includes a combinatorial rule that describes how different mixes of levels of person and event variables combine in determining the degree to which negative threat appeals will persuade people to act. Research on preparation for earthquakes (Mulilis & Lippa, 1990) and tornados (Mulilis, Duval, & Bovalino, 2000) has provided support for this model. Another approach to understanding interpretive processes and their relationship to preparedness derive from how people's discussion of hazard issues with others represents the trigger for the hazard preparedness process.

Critical Awareness

Drawing on earlier work investigating how people adapted to volcanic eruption consequences (Paton et al., 2001), Paton et al. (2005) developed a theory of hazard preparedness that derived its name from the use of the construct of critical awareness as a foundation for understanding how people prepare for hazard events. Critical awareness (Dalton, Elias, & Wandersman, 2001; Seedat, 2001) describes the extent to which people think and talk about a specific source of adversity or hazard within their environment.

According to Dalton et al. (2001), the critical awareness construct taps into the degree to which people perceive that issues are critical or salient for them, with this predicting the degree to which people perceive a need for action. This made it a promising candidate as a variable that could further understanding of the degree to which people would be motivated to prepare for infrequent events. That is, hazard preparedness is a process motivated initially by the salience people attribute to it, and this salience is, at least in part, socially constructed (Paton et al., 2005). The key variables and the relationships between them are summarized in Figure 5.5.

The Critical Awareness theory also allocated a role for risk perception (Johnston et al., 1999; Lindell & Perry, 1992; Weinstein et al., 2000) and anxiety as variables capable of motivating hazard preparedness. Anxiety about natural hazards (e.g., earthquakes) can influence people's decisions about whether or not to adopt protective measures (Dooley, Catalano, Mishra, & Serxner, 1992; Duvall & Mulilis, 1999; Lamontagne & LaRochelle, 2000; Marks & Matthews, 1979; Paton, Buergelt, & Prior, 2008). Anxiety can also increase the likelihood of people transferring responsibility for their safety to others (Weinstein, Lyon, Rothman, & Cuite, 2000) and, therefore, not preparing.

The Critical Awareness Theory proposed that the relationship between motivation to act and intentions to prepare was mediated by outcome expectancy (also called response efficacy), self-efficacy, and action-coping variables. Finally, the theory included variables that could moderate the conversion of intentions into actual behavior. These include the personal capabilities and resources (e.g., time, skill, financial and physical resources, social network availability, and competencies), trust, willingness to accept personal

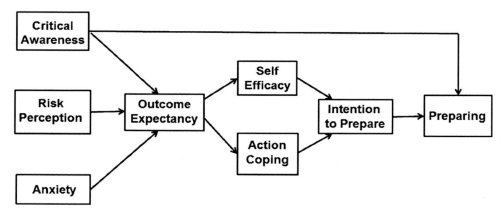

Figure 5.5. The Critical Awareness Theory (adapted from Paton et al., 2005).

responsibility for their safety (Duval & Mulilis, 1999; Lindell & Whitney, 2000; Mulilis & Duvall, 1995; Paton et al., 2000) and beliefs regarding when the next hazard event would occur (Dooley et al., 1992; Mulilis & Duvall, 1995).

Analysis confirmed that while risk perception did play its expected role as a precursor of preparedness, it was a poor predictor of either behavioral intentions or actual preparedness (Paton et al., 2005). The roles of Critical Awareness and Earthquake Anxiety as precursors were also supported. Consistent with the theoretical predictions derived from this theory, the influence of these motivating factors was mediated by outcome expectancy and self-efficacy, with outcome expectancy preceding efficacy and coping judgments (Figure 5.5).

The analysis of relationships between intention and preparing identified moderating roles for trust and the resources (e.g., time, skill, financial) required to plan and implement strategies. The expected timing of the occurrence of the next damaging earthquake also moderated the conversion of intentions into action. While respondents who believe an earthquake could occur in the next 12 months were highly likely to convert intentions into preparedness, those who anticipated its occurrence beyond 12 months were significantly less likely to do so. This theory reiterates the importance of self-efficacy as a predictor of hazard preparedness. This variable also plays a prominent role in recent iterations of another theory; the Theory of Planned Behavior.

Another common denominator between the Critical Awareness and Planned Behavior theories is their inclusion of a social influence on action. In the Critical Awareness theory, this is in the form of the frequency with which people discuss hazard issues with others. This is more likely to arise if preparedness takes on the guise of a social norm. Normative influence is identified as a pivotal predictor of behavior in the Theory of Planned Behavior. Taking these elements together, it becomes more apparent that preparedness is influenced by how people engage with and relate to others.

The Theory of Planned Behavior

The theory of planned behavior proposes that behavior is a product of intentions that are predicted in turn by three factors: people's attitude toward the target behavior, their "perceived subjective norm" (which includes their judgments about social pressures to perform an action), and their perception of behavioral control (and in some later versions, self-efficacy), which refers to people's perception of how difficult it is to perform the target behavior (Ajzen, 1991). The theory argues that people's response to a situation (e.g.,

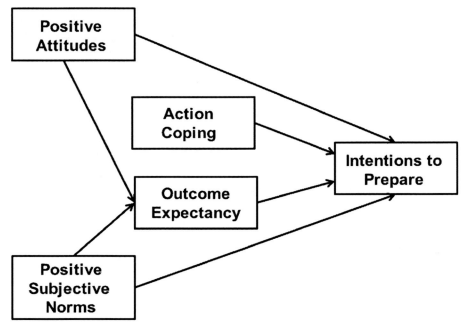

Figure 5.6. Adapted version of the Theory of Planned Behavior (adapted from Ajzen, 1991 and McIvor and Paton, 2007).

their preparation for a hazard) is affected more by their beliefs about the effectiveness of a given behavior, than by their beliefs about the hazard that warrants action (Figure 5.6). This theory has been moderately successful in enhancing health behaviors (e.g., smoking cessation) and the adoption of preventive (e.g., use of helmets) behavior (Armitage & Conner, 2001; Hardeman et al., 2002). It has also been applied to earthquake preparedness (Lindell & Perry, 2000; McIvor & Paton, 2007). This theory highlights how the attitudes people hold regarding hazards, their mitigation, and the social norms that influence the acceptance or rejection of preparedness actions will influence preparing (Bagozzi, 1992; Bennett & Murphy, 1997; Donald & Cooper, 2001).

The attitudes a person holds towards a particular behavior are influenced by the positive or negative consequences they perceive as resulting from performing the behavior (Doll & Ajzen, 1992; Park, 2000). Individuals can hold any number of attitudes, but they are limited in the number they can attend to at any one time. Attitudes are organized in a hierarchical fashion according to the relative salience they have for a person. The greater the salience of a particular attitude, the more likely it is to influence a person's intentions and behaviors (Bagozzi & Dabholar, 2000; Doll & Ajzen, 1992). Thus, while a person may hold a positive attitude to hazard preparedness, if they attach greater salience to other issues such as crime, health, and education, their hazard attitude is less likely to motivate action (Doll & Ajzen, 1992; Flynn, Slovic, Mertz, & Carlisle, 1999; Paton, 2003–see also Chapter 4). The infrequent nature of hazards can reduce the salience of hazard attitudes relative to those relating to other aspects of people's lives.

Subjective norms also influence behavioral choices (Doll & Ajzen, 1992). Subjective norms reflect beliefs concerning the social expectations of significant others (parents, spouses, friends, peer group, etc.), and an individual's compunction to align their behavior with the expectations of these significant others (Park, 2000; Smith & Terry, 2003). If a person believes that their significant others hold favorable attitudes towards a particular behavior, or if they believe that performance of a specific action would be viewed favorably by them, they are more likely to perform that behavior (Doll & Ajzen, 1992; Park, 2000; Smith & Terry, 2003).

McIvor and Paton (2007) tested a highly adapted version of the TPB for earthquake preparedness. They found that positive attitudes had direct and indirect influences on intentions to prepare (Figure 5.6). The influence of positive attitudes and positive subjective norms on intention to prepare was, as predicted, mediated by outcome expectancy. Action coping had an independent influence on intentions to prepare only. Positive subjective norms had no direct influence on intentions to prepare, but had an indirect influence mediated by outcome expectancies. Positive subjective norms were positively related to people's belief that protective action would be effective in managing hazard consequences. The results reiterate the fact that what people do to manage their risk is influenced by the degree to which significant others (e.g., family, friends) endorse or reject actions (Carroll et al., 2005; Gordon, 2004; Prewitt Diaz & Dayal, 2008; Proudley, 2008).

McIvor and Paton's (2007) findings support the contention that positive attitudes to hazard mitigation and positive social norms about hazard preparedness increase the likelihood that people will adopt protective measures for earthquakes. Attitudes are difficult to change, and more so when the target of the attitudes relates to infrequent events such as earthquakes. One approach that could be adopted would be to invite people to first produce reasons why preparing could be effective (for them, their family, their house, their employment, and so) and only after they have done so, provide them with information on what to do. Behavior change in the direction of increased preparedness will be more likely if ideas about preparedness and its benefits are endorsed by the significant others in a person's life.

Significant others (family, friends, etc.) who view favorably the adoption of protective measures can exercise mutual influence on each other's beliefs regarding the efficacy of preparing for natural hazards. Because it can foster and re-affirm these beliefs, engaging in discourse with significant others plays an important role in developing and, particularly, sustaining preparedness (see also Critical Awareness theory above). The more people actively discuss hazards and the protective measures that can be implemented to mitigate their consequences, the more salient hazard issues will become in a community. This, in turn, facilitates the development of attitudes and subjective norms conducive to the adoption of protective measures (Hardin & Higgins, 1996). Thus it would appear beneficial to invite people individually and collectively to identify reasons why preparing can be effective and ensure that strategies to encourage discourse about natural hazards and protective measures target communities, groups, families, and friends, rather than focussing risk communication on individuals (Doll & Ajzen, 1992; Flynn et al., 1999; Marris, Langford, & O'Riodan., 1998; Paton, 2005).

Further research is needed to examine how hazard attitudes are formed, sustained, and organized, as well as how they can be changed to facilitate the sustained adoption of protective measures. Work also needs to be directed to identify those with whom normative comparisons are made as well as the relative influence of different referents. That is, can referents (significant others) be differentiated with regard to their relative influence on hazard preparedness decisions? Are some referents more influential than others? For example, Trafimow and Fishbein (1994) drew a distinction between general and specific referents. Referents such as parents and spouses were important irrespective of the behavior under consideration. This would suggest that family dynamics would represent an appropriate target for intervention intended to increase positive normative influence. This view is echoed in the social support literature that suggests that the effectiveness of emotional and informational support is a function of who provides it; family for the former and experts for the latter. However, other referents may have greater credibility when discussion turns to the more technical aspects of hazard processes and preparedness.

For example, a geologist or risk management referent may be appropriate when the focus of intervention is on the relationship between hazards and effective mitigation (see Chapters 2 and 3). A further consideration is whether sections of the community can be differentiated with regard to the source(s) they find most influential (Latimer & Martin-Ginis, 2005). When risk communication functions as a top-down process that disseminates information to people, this issue may not seem important. However, as risk communication moves to become a process that involves stakeholders engaging with others to develop their complementary roles in risk management, an understanding of the relative contribution of different referents becomes more significant.

If groups can be differentiated with regard to their relative salience as referents, strategies will need to be developed to accommodate this aspect of diversity (Paton, 2005). For example, people often prefer to receive information about hazards and hazard preparedness from those in similar circumstances to themselves (e.g., information from people in other communities who comment on the effectiveness of mitigation to deal with hazard consequences—see Chapter 4). Information from community sources tends to be more effective than information from formal sources, with issues such as perceived similarity, sense of shared fate, and trust being influential in this process (McIvor, Paton, & Johnston, 2009; Paton, 2008). However, relationships with formal, expert sources become more important when people recognize a need for technical input into their decision making.

The theoretical approaches discussed so far focused on identifying the social-cognitive competencies that influence how people make choices. It is also important to consider how people acquire and use information. The two remaining theories discussed in this chapter, social marketing and the Protective Action Decision Model, focus on how people interact with sources of information to acquire the information required by them to make their decisions. The first, social marketing, adopts some of the approaches introduced above (e.g., cost-benefit issues) but focuses on marketing techniques to present information.

Social Marketing

Social marketing is increasingly being used to guide the development of hazard public education programs (Faulkner & Ball, 2007; Evans, 2006; Hastings & McDermott, 2006; Smith, 2006; McKenzie-Mohr, 2000; McKenzie-Mohr & Smith, 1999; Bloom & Novelli, 1981; Kotler & Zaltman, 1971). Social marketing approaches aim to both reduce behaviors that increase risk, and encourage the adoption of actions that reduce risk or offer better protection or increased safety (O'Neill, 2004). Social marketing uses commercial marketing techniques to plan, execute, and evaluate programs that are designed to influence the voluntary behaviors of citizens to improve their personal welfare and that of their society (Andreason, 1995; Kotler & Zaltman, 1971). Its effectiveness in this regard is attributed to its being based on research about behavior change, it focusing on the needs and wants of the "consumer;" and its emphasis on evaluation.

As a strategy, social marketing proposes a sequence of activities (McKenzie-Mohr, 2000). The first step is to target behaviors that have a substantial effect on outcomes, such as strengthening house foundations in the case of earthquakes. The second step is identifying barriers to the desired behavior (e.g., internal barriers such as fatalistic attitudes and external barriers such as resource restraints). Surveys and focus groups are used to identify these barriers, and strategies to surmount them are then identified. For internal barriers, techniques such as "foot in the door" strategies building on initial small commitments can be used (see also Chapter 3), whereas for external barriers, cross-disciplinary collaboration is normally required. McKenzie-Mohr (2000) cited case studies showing how these strategies have been effective in changing environmental behaviors.

Social marketing often uses dramatic imagery. While fear-inducing imagery or communication provides graphic illustrations of the issues people could have to contend with during a hazard event, it can backfire if it instills or strengthens people's belief that the magnitude or intensity of events means that they cannot do anything to mitigate their risk (see Chapter 4). That is, dramatic imagery can increase anxiety to the point where it acts to reduce the likelihood of people taking action to prepare themselves (e.g., Paton et al., 2005).

Fear-inducing communications can enhance short-term reductions in harmful behaviors, but their effectiveness tends to diminish over time (Daniel, 2007a). Consequently, some researchers have questioned the use of fear messages in social marketing. Hastings, Stead, and Webb (2004) observed that although controlled laboratory studies have found positive effects for fear messages, such findings have not generalized well to real-world settings, as people may switch off the source of the communication or become cynical or hardened to it. In addition, if fear messages are clearly targeted at specific groups, other groups can infer that they are not at risk and become more complacent. This point, of course, applies not only to fear messages. Hastings et al. propose that messages such as humor and irony may be more effective than fear messages, especially with teenagers and young adults (see also Sharpe, 2009).

Social marketing has been criticized as being manipulative (Grier & Bryant, 2005; Kotler & Zaltman, 1971), placing more emphasis on persuasion than informed decision-making (Evans, 2006; McKenzie-Mohr & Smith, 1999; Morgan et al., 1992), and being inadequately evaluated (Grier & Bryant, 2005; Smith, 2006). While identified as a viable mechanism for promoting behavior change (Gordon et al., 2006), questions remain about its ability to do so when the targeted actions have to be repeated over time or where the behavior change required is complex (Bloom & Novelli, 1981), such as with natural hazard mitigation.

A development of the social marketing approach, Community-based (C-B) social marketing focuses on: identifying barriers and benefits to behavior change, identifying behavior change tools, conducting pilot studies, and program evaluation followed by review and revision (McKenzie-Mohr, 2000;

McKenzie-Mohr & Smith, 1999). C-B social marketing can be differentiated from mainstream social marketing by its greater use of psychological theories of barriers to behavior change in the design of interventions. C-B social marketing requires the information given to recipients being broken up into segments, with communication tailored for each segment (Evans, 2006; Smith, 2006; McKenzie-Mohr & Smith, 1999; Bloom & Novelli, 1981—see also Chapter 3). Another well-researched and important approach involving segmenting of information is the Protective Action Decision Model.

The Protective Action Decision Model

The Protective Action Decision Model (PADM) argues that people's decisions about how to respond to a perceived threat depend upon the judgments they make about three factors; the threat posed by a hazard, identifying the actions available and capable of protecting them against hazard consequences, and identifying the sources from whom information about hazard and protective measures can be obtained (Figure 5.7). The PADM proposes that this involves people proceeding through several stages of information seeking and evaluation: detection/warning, psychological preparation, logis-

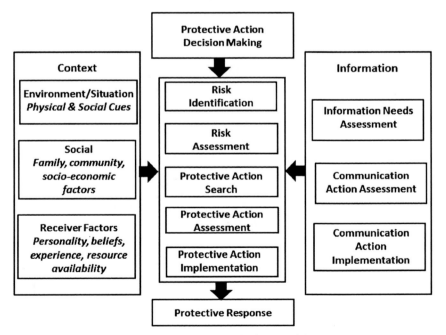

Figure 5.7. The Protective Action Decision Model (PADM) (adapted from Lindell, Prater, & Perry, 2006; Tierney, Lindell, & Perry, 2001).

tical preparation, and protective action selection/ implementation (Lindell & Hwang, 2008; Lindell & Perry, 1992). The information required for people to act at each stage can be sourced from the physical environment and/or people's social context (e.g., other people in their community, emergency management, media, etc.).

The PADM places considerable emphasis on how uncertainty influences both the search for information and the nature and quality of protective action decisions. Lindell and Perry (1992) propose an inverse relationship between the uncertainty and ambiguity present in the decision context and people's compliance with recommendations contained in risk communications. They also make the important point that as ambiguity and uncertainty increase, people spend more time seeking and processing information. In doing so, they are less likely to prepare for and implement appropriate protective actions and may delay or abandon actions. The degree to which the information available to people allows them to negotiate this uncertainty influences their progress (or whether progress takes place at all) through the protective action, decision-making process. In this context, it is important to appreciate that people's appraisal of whether information can resolve their uncertainty can also be affected by their perception of or beliefs about the source, particularly with regard to how their past experience with a source has influenced the degree to which they trust a source (Paton, 2008—see also Chapters 6 and 7).

An important facet of the PADM is its highlighting how it is the interaction between information and people's competencies in information management and decision making (e.g., skills in information search and problem solving) that predicts action and not just information alone. The PADM subdivides the process of determining protective action adoption into two phases: the pre-decisional and the decision-making stages.

Pre-decisional Processes

Prior to engaging in protective action decision making per se, people must first become aware of a need for them to do so. Whether people become aware of this need is a function of two things. The first concerns the quality of people's exposure to, attention to, interpretation of, and comprehension of environmental cues. The second, either on its own or in conjunction with the former, is their receiving, attending to, and interpreting and comprehending information from their social context that identifies the existence of issues in their environment that requires some response from them.

The PADM identifies environmental cues and socially-sourced data as being independent. People may thus fail to progress from a pre-decisional stage if they lack a pertinent level of environmental awareness. For example, Paton, Buergelt, and Prior (2008) noted that people's awareness of meteorological and environmental signals of wildfire risk influenced preparedness, but not everyone (even when living in the same location) possessed a level of environmental awareness commensurate with their understanding how their risk changed as environmental conditions change. Accordingly, some people thus failed to use environmental cues as triggers for action. Similarly, differences in social and interpersonal skills and people's knowledge of community sources from which pertinent information could be obtained introduced variability in people's ability to access and use information from their social context (Paton, 2008). Only when information from social and physical environments can be identified, accessed, and interpreted in ways that meets people's needs and expectations and reduces their uncertainty is it capable of being used to make decisions and guide actions (Lindell & Perry, 2004; Paton, 2008).

The PADM thus offers insights into why some people do not prepare, and offers ways of accounting for differences in both the level of adoption and the nature of the items adopted (see Chapter 3). If, however, people attend to environmental and social data and interpret these data in ways that identify a hazard whose existence has implications for them, they can progress to the decision stages in the model.

Decision Stages

If people's risk information and/or environmental cues create awareness of hazard issues, the PADM proposes that people then enter a sequential, multi-stage, decision-making process. These stages are: risk identification, risk assessment, protective action search, protective action assessment, and protective action implementation. In parallel with these, the PADM includes information search and evaluation functions (assessing information needs, identifying pertinent sources of information, and assessing when information is needed) that inform all stages of the decision making process (Figure 5.7).

Accepting the existence of a hazard capable of posing a threat to them does not automatically stimulate people to action. For action to occur, people have to evaluate and personalize the consequences a hazard event could have for them. If this risk assessment affirms susceptibility to experiencing adverse hazard consequences, this elicits *protection motivation.* If people believe that a threat exceeds an acceptable level of personal risk, they move to the next stage: *protective action search.*

According to the PADM, protective action search triggers recalling protective actions (what could be done to protect themselves and their property) from their past experience or from other sources with pertinent knowledge or experience. This process results in the generation of a *decision set* that comprises possible protective actions. The contents of the decision set include options from several sources (e.g., family, friends, neighbors, others within the communities in which they interact, people they work with, media sources, and emergency management sources). This process can provide people with a pool of potential adjustments, all of which could be adopted. It is rare for people to adopt all possible adjustments. Instead, they select from the options available to them (see Chapter 3) in a procedure labeled *protective action assessment* (Lindell & Whitney, 2004).

Protective action assessment involves people assessing alternative actions, evaluating their implications compared with not acting or not altering their current actions, and deciding which actions are most appropriate for them given their needs and circumstances. The selection of appropriate protective methods culminates in the development of an *adaptive plan.* However, this selection need not be accomplished through the objective assessment of needs in relation to hazard consequences. Rather, several factors influence what people decide to incorporate into their plan.

Prominent influences on what people decide to do are their assessment of the potential for a measure to effectively increase safety and/or reduce the likelihood of their experiencing adverse effects from hazard activity (cf. response efficacy or outcome expectancy in the HBM, PMT, PrE, and critical awareness theories) and the perceived *time requirements* (people's assessment of the number of steps required for implementation and how long they are likely to take) for implementation (Lindell & Perry, 2004). To this, Lindell and Perry add *perceived implementation barriers* (e.g., levels of knowledge, skill, tools and equipment, or social cooperation required to achieve protection), out-of-pocket expenses (e.g., fuel), opportunity costs (e.g., employing someone to secure a house to its foundations versus using these funds to purchase a needed household appliance), and aesthetic cost (e.g., the unattractive appearance of houses that are elevated out of the flood plain) to the list of potential constraints on action. They also point out that the higher people perceive these costs to be, the more likely they are to abandon their plans or delay implementation until they are certain it is necessary. A similar finding was reported by Paton, Buergelt, and Prior (2008).

Developing a plan is important, but the ultimate utility of any plan is a function of its implementation. The final stage, *protective action implementation,* assumes that people exposed to a specific hazard have determined that action on their part is required, they have selected at least one potentially effective measure from those they identified as being capable of offering them protection from hazard consequences, and they have adopted those that they can (within the context of their resource and competency limits). Implementation can be delayed by not wanting to divert resources (e.g., money, time) from more personally interesting or pressing (e.g., going on holiday, buying a new TV, need for household repairs, etc.) goals and the low salience of hazard issues (e.g., compared with problems such as crime, health care, etc.). Consequently, implementation decisions may be delayed until a threat of sufficient magnitude to warrant disruption of normal activities stimulates adoption decisions (see also Paton, Buergelt, & Prior, 2008).

Information Search and Evaluation

To progress through these stages, people need information. Consistent with the social-cognitive models discussed above, the PADM acknowledges that people are not passive recipients of information, but actively interact with information sources in ways that reflect their needs and expectations. To meet these needs, the PADM includes three information access processes that operate in parallel with the decision stages.

The first of these, *information needs assessment,* involves people formulating questions that culminate in an *identified information need* (regarding hazards, hazard adjustments, and alternative protective actions that match their beliefs regarding their needs and expectations). The second process that informs decision making concerns *communication action assessment.* This addresses the fact that identifying one's information needs and knowing where and from whom to obtain information that meets one's needs are not the same. Resolving this issue leads to selecting both information sources and channels/media in the context of their developing an *information search plan.* The sources from which information is sought differ depending upon the stage of the protective action decision process that has generated the need for information (Lindell & Perry, 2004). For example, information about hazards themselves can prompt questions being directed to emergency management or risk management experts (see also discussion of referents above). In contrast, to acquire information on preparedness actions that are consistent with their circumstances and needs, people may turn to those in similar circumstances (e.g., neighbors, people they socialize with) and with whom they identify (Paton, 2008).

The final information search process is *communication action implementation,* which provides *decision information.* The dominant driver of this component of the process concerns people's beliefs regarding the time frame within which action is required. If the need is immediate, people are more likely to actively seek information from the most appropriate source through the most suitable channel (Lindell & Perry, 2004; Mulilis & Duvall, 1995). People are, however, less likely to do so if they believe the event is unlikely to occur until sometime well into the future (Paton et al., 2005).

Communication action implementation leads to one of several outcomes. If a query elicits information that answers the question that prompted the search for information, then the information seeking component of the process is successful and facilitates advancing the protective action decision process. If, however, the source is unavailable or fails to offer meaningful information, uncertainty is not reduced and this can delay the protective action decision process or lead to its termination (Eng & Parker, 1994; Lindell & Perry, 2004). This reiterates a need to consider the level of competence (e.g., problem definition, problem solving) that people bring to this process. Information may be available, but if people do not know how to formulate their information needs, instigate a search for answers, or represent their needs, they may be unable to source information or use what is available effectively (Paton, 2008). This is not because information is unavailable, just that people do not know how to access it (in the context of having to make atypical decisions about complex events characterized by uncertainty—see also Chapter 4). On the other hand, people who have the requisite information search and decision-making skills may be thwarted in their protective action decision making if they interact with agency sources that only make general data available (i.e., that does not answer their specific questions). These issues are explored further in Chapters 6 and 7 in the context of discussing empowerment.

The social cognitive theories discussed in this chapter highlighted the fact that effective risk communication requires more than just providing high quality information to people. Effective risk communication also entails ensuring that people have the individual (e.g., self-efficacy) and social (e.g., norms) competencies and characteristics that influence their ability to access, interpret, and use information to appreciate the need for preparing and to advance their state of readiness. While focusing primarily on the analysis of how people make their decisions, several theories discussed in this chapter alluded to how social relationships (within communities and between people and civic agencies) could be implicated as predictors of preparedness. This introduces the need to expand the scope of the discussion of preparedness to include social relationships and social contexts in the conceptualization of preparedness. This issue is picked up in Chapter 6.

Chapter 6

SOCIAL INFLUENCES ON HAZARD BELIEFS

INTRODUCTION

The Critical Awareness Theory, the Theory of Planned Behavior, and the Protection Action Decision Model that were discussed in the previous chapter introduced how certain characteristics of people's relationships with others in family and community contexts and their social behaviors (e.g., social norms, discussions of hazard issues, etc.) influenced preparedness. These works illustrate only the tip of the social influence iceberg. This chapter seeks to expand understanding of how people's social (e.g., those with whom people interact on a regular bases) and societal relationships and the social exchange processes that occur in social interaction guide people's thinking about hazards and risk and affect what they decide to do to manage their risk (Dake, 1992; Dow & Cutter, 2000; Lasker, 2004; Marris et al., 1998; Paton, 2008; Rippl, 2002; Rohrmann, 1994; Shinn & Toohey, 2003).

The studies mentioned in the previous paragraph introduce the important role that relationships between stakeholders play in risk management. To this can be added a need to consider how *interdependencies* between stakeholders contribute to the development of comprehensive risk management. Drawing a distinction between interrelationships and interdependencies, and seeing both as playing roles in risk management, signals a shift in how risk management is conceptualized. It describes a shift from simply acknowledging that stakeholders become related (either directly or indirectly) in the processes of developing and implementing risk management plans (e.g., risk management agencies develop and disseminate information to stakeholders) to one in which all stakeholders, including community members, are seen as sharing responsibility (albeit not to the same extent) for risk management and as playing complementary roles in how comprehensive risk management plans are developed and enacted.

For example, societal sources reduce risk through creating building codes and the inhabitants of building (domestic and commercial) complement this contribution to public safety by reducing risk through actions such as securing furniture and fittings to the walls of their home or securing office equipment and computers. However, citizens may have little influence on building code specifications and risk management agencies cannot compel people to secure furniture and fittings. But if each of these stakeholders plays their part, their actions complement one another and contribute to a greater level of public safety than would have occurred if each acted alone.

Thus, when stakeholders undertake activities that complement one another, their actions contribute to more comprehensive risk management. While such interrelationships have often been implicit in the past (e.g., agencies assume that people are acting in the ways they expect them to), a shift to the more engagement-based social risk management makes this more explicit (see also Chapter 7). The importance of examining risk management in this way derives from the infrequent and complex nature of the hazard events people need to prepare for.

In the absence of being able to find out about hazard phenomena for themselves, people turn to others to help them make sense of hazards and what they can do to manage the risk they pose. These others can be members of people's communities, but they can also be the civic agencies that develop and disseminate risk information and the media agencies that interpret this and other information on hazards, risk, and disasters to the populace at large. This chapter explores these issues from three perspectives (summarized in Figure 6.1).

The chapter first introduces a theoretical paradigm: symbolic interactionism. This paradigm captures how the meaning (i.e., from how they develop their mental models—see also Chapters 2 and 4) people attribute to environmental events and how they could be managed is constructed through social interaction. Using this paradigm to inform understanding an engagement-based focus is important for two main reasons. The first relates once again to the fact that the infrequent nature of complex hazard events makes people more reliant on others for the information used to construct their beliefs about risk and its management. The significance of this is discussed in this chapter in relation to how societal (e.g., civic agency, media) institutions influence people's risk beliefs. In particular, it emphasizes how the kind of filtering and processing the media apply to hazard and risk information can adversely impact on people's risk beliefs and whether they prepare. That is, the raw material for the construction of people's beliefs can be distorted and create mental models (see Chapter 2) that are more difficult to challenge. The only way to do so is to engage with people in ways that challenge these

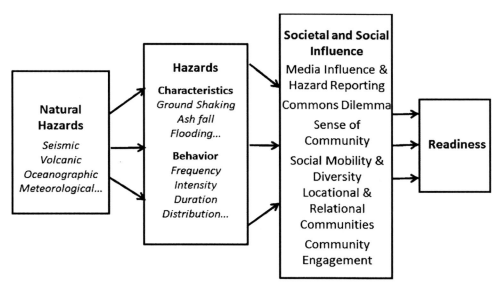

Figure 6.1. Societal and social influences on preparedness.

media-developed ideas and provide scope for checking and changing errors and misconceptions.

The second reason why an understanding of the social construction of hazardous realities is important stems from the fact that understanding is socially constructed. This introduces a need to critically evaluate the nature of social network characteristics in contemporary societies. Consequently, this chapter concludes with a discussion of how the changing nature of social relationships in contemporary neighborhood and community contexts may be reducing the opportunities to appraise risk information available to people in urban environments. Both of these processes have important implications for understanding the social context in which risk management occurs. The chapter starts with a discussion of how people's relationships and social interactions contribute to how they construct their reality.

PREPARING IN THE CONTEXT OF A SOCIALLY CONSTRUCTED REALITY

Symbolic interactionism (Blumer, 1969) argues that people actively and constantly interpret stimuli from the social and physical environment while interacting with the elements in that environment. People integrate these interpretations through a process of reflection using pre-existing mental models (see Chapters 2 and 4). It is through this process of reflecting on experience that people construct the meaning of the things they interact with.

The socially constructed understandings that guide behavior develop through everyday conversation and the transactions and interactions people have with each other and with the environment they share and interact in. The meanings developed through interaction make significant contributions to how people shape, construct, perceive, and interpret their shared reality (e.g., Berger & Luckmann, 1966; Burr, 1995; Gergen, 1985). These representations also guide how people communicate their experiences and the meanings they impose on them when talking with others and when engaging in discussion with others (Deaux & Philogene, 2001; Flick, 1998; Moscovici, 2000). These constructionist processes apply as much to how people interpret and prepare for hazards as they do to the other aspects of people's everyday lives, including how people conceptualize risk and make choices about how to deal with it (Paton, Buergelt, & Prior, 2008).

This process results in people constructing "group-specific" meanings for the objects, events, and relationships they encounter in their lives. Consistently interacting with people with shared interests or goals (e.g., in workplaces, social meetings, through community memberships, etc.) contributes to the development of interpretive frameworks that are shared within the groups or communities in which people regularly interact. Once meaning is derived, people then act towards the objects, events, ideas, and relationships to which meaning has been attached in ways consistent with these meanings (Blumer, 1969). These are not static. Interpretations and their associated actions constantly evolve, with this process defining how people adapt to new conditions (Denzin, 1992). The ultimate function of this interpretive process is to facilitate people's ability to adapt as well as possible to changes in the environment.

Just as with other phenomena in their lives, people's social representations of risk and risk management phenomena are thus shared, collectively elaborated, and evolving assumptions about their physical and social worlds that provides them with a framework for interpreting and making sense of risk issues. However, while this process works well (e.g., with regard to developing shared meanings between groups) for the more mainstream events that people encounter (e.g., because they share experiences or have had common experiences), natural hazard mitigation poses a more challenging set of circumstances for this meaning-making process.

A significant contribution to this challenge arises from the infrequent nature of the events people are being asked to prepare for. While interaction with other community members still acts to influence how risk beliefs are developed and enacted (see below and Chapter 7), the rarity of hazard events and their complexity means that the experts (e.g., scientists, risk management professionals) who provide information about hazards, risk, and how risk can be managed become important influences on people's social construction of hazard events and risk.

A further challenge can be discerned here because much of the interaction that occurs with expert sources of information is indirect (e.g., via public education materials delivered via booklets, DVDs, etc.). In the absence of opportunities to discuss issues and, more importantly, check on the meanings people derive from risk information, then the kinds of misconceptions and misunderstandings discussed in Chapter 2 can become more permanent components of people's risk beliefs. The importance of the latter is amplified by the fact that, in a risk management environment, the information available to people and which they use to make sense of hazard phenomena often reaches them after being filtered and interpreted by civic agencies and, particularly, by the media (including their reporting on and interpreting the causes, consequences, and issues that arise in other disasters).

CIVIC AND MEDIA INFLUENCE

The views of the civic risk management experts presented to the public are socially constructed to some extent. For example, the information made available to the public can have been filtered in ways that reflect institutional values, societal or agency views about the functions risk information is used to pursue (e.g., providing objective knowledge versus presenting information in ways that minimize mitigation expenditure), and by political and economic considerations (see Chapter 2). However, to fully appreciate how this social filtering process can affect citizen's risk beliefs, it is particularly important to consider the role of the media (Carvalho, 2007; Danesi, 2002; Olson & Rejeski, 2005; Pettenger, 2007; Taylor, 1983; Weinstein et al., 2000).

Media coverage and interpretation of hazard events and disasters is often selective and sensational and filtered in ways that amplify or attenuate risk (e.g., Flynn, Slovic, Mertz, & Carlisle, 2001; Kasperson, Renn, Slovic, Brown, Emel et al., 1988; Pidgeon, Kasperson, & Slovic, 2003; Sjöberg, 2007). Information about risk that is amplified by a selective focus on the more sensational and destructive aspects of a disaster has significant implications for how people develop their risk beliefs and whether these beliefs translate into action (see also Chapter 4).

Sensationalizing Hazard Reporting

Reports of disasters such as earthquakes in the news media are typically sensational. However, reports and feature articles vary in their style and serve different functions. In particular, news media reports issued sometime after an earthquake has occurred differ from those reported immediately after the earthquake. Cowan, McClure, and Wilson (2002) compared the effects of news reports written immediately after the 1995 Kobe earthquake with articles written one year later ("anniversary" articles). Articles written immediately after the earthquake reported widespread damage and emphasized the role of the earthquake in producing the damage, with headlines like: "Earthquake ravages Kobe." Reports written a year later contrasted the design of damaged and undamaged buildings and focused on lessons that could be learned from the earthquake, with headings like: "Lessons from Kobe." When these reports were presented to participants (with references to Kobe removed), the "year after" reports led people to attribute the damage to building design more than the "week after" reports. This finding suggests that articles that analyze damage in earthquakes lead to less fatalistic views than the typical "catastrophe" reports written immediately after an earthquake. Thus media reporting that focuses people's attention on the hazard events (i.e., the earthquake) rather than on more mutable characteristics such as building characteristics contribute to creating a more complex risk management environment, and one which is more likely to discourage people from preparing.

Recent research on media reports of the 2011 Christchurch earthquake showed that both fatalistic and informed messages were communicated in the media in the weeks directly after the earthquake, not just much later (Velluppillai & McClure, 2012). This study found examples of both types of reports in the month after the Christchurch earthquake. An example of a fatalistic excerpt is: "The fact of the matter is that no building in the world will hold up if you've got this sort of ground movement. You can have the best architects, the best engineers and the best contractors, but if nature's going to drag things away from the foundations, there's nothing you can do." An example of an informed excerpt is: "NZ requirements for earthquake design have been progressively upgraded since 1935. With some exceptions, old buildings performed poorly and new buildings came through well, especially given the extreme shaking." The study presented composites of these messages to two groups of participants, who then judged how much the damage to buildings was caused by building design and earthquake magnitude and how much it could have been prevented.

The results showed that despite citizens' extensive real-world knowledge about the Christchurch earthquake, the different media messages had different effects on judgments, with those reading more informed reports attributing the damage to building design and seeing damage as more preventable than those reading fatalistic reports. This was despite the fact that both groups judged the amount of damage in the earthquake to be the same. This demonstrates how the media can have either positive or negative effects on citizens' fatalism about earthquakes and other hazards. This illustrates how media reporting can result in different social constructions of hazardous realities. Acting on these social constructions can result in people being more or less motivated to prepare. Thus, media coverage can discourage preparedness and create a more complex risk management environment in the future. This highlights how interdependencies exist between civic risk management agencies and the media. How these institutions engage with one another, and with the communities they serve, will thus play a significant role in developing comprehensive and effective risk management. The importance of considering this issue can be further illustrated with regard to considering how people's social construction of their risk and their motivation to prepare is also strongly influenced by how damage is reported by media sources.

Distinctive Versus Generalized Damage

News media usually present pictures of generalized damage, rather than the distinctive damage that is more common with earthquakes in regions with reasonable building standards, such as Japan and the USA. When natural disasters occur, news media focus on scenes where the most damage occurs (Gaddy & Tanjong, 1987; Hilton, Mathes, & Trabasso, 1992; Hiroi, Mikami, & Miyata, 1985). News reports also describe disasters in ways that accentuate the magnitude and severity of damage. These portrayals convey implicit causal models (i.e., beliefs regarding the causes of specific outcomes) of earthquake damage. This point is illustrated by comments from an American journalist arriving in Kobe after the earthquake in 1995. "I was amazed how much of Kobe was still there. . . . I mean, I had watched hours and hours of TV in America about this earthquake, and I had no idea that there were houses and tall buildings still standing all over the city" (International Herald Tribune, February 2, 1995, p. 2).

The pattern of information presented in news media can increase fatalism and lead people to attribute damage to the earthquake (over which control is not possible) rather than building design (which is amenable to control). The research described here suggests that fatalism would be reduced if news media showed that earthquake damage in countries where building standards are applied is often distinctive. The same applies if media portray buildings that stand firm because of adjustments made in preparation for earthquakes (e.g., base isolation). It may not be possible to change the media, but risk communication and public education strategies can portray the patterns of damage in ways that lead people to see that this damage is caused not only by the earthquake but also by the design of the buildings, soil composition, and other infrastructure.

In addition, research has shown that scenarios that describe the proportion of buildings with different designs that are damaged leads to lower levels of fatalism and to controllable attributions for the damage. Research in attribution theory demonstrated that people's attributions for events revolve around a search for the causal mechanisms that leads to different outcomes (Ahn & Bailenson, 1996). In relation to disasters, these mechanisms include the design characteristics that affect the level of damage created in a disaster. Knowledge of these processes is important not only for understanding how the media present information, but also for how engineers and risk management experts convey information to the public.

Applying this "causal mechanism" principle to attributions for disaster outcomes, research examined the effects of scenarios where engineers linked the damage to specific buildings in earthquakes to the design of those buildings (McClure, Sutton, & Wilson, 2007). This study examined how these messages affected attributions for earthquake damage and judgments that the damage could be prevented. The results showed that scenarios that described the faulty design of buildings that suffered more damage in earthquakes led to less fatalistic attributions for earthquake damage than scenarios that omitted this information. The design information had particularly strong effects when it concerned the structural weakness of buildings that were damaged rather than the resilience of buildings that were undamaged. This difference may reflect people's greater sensitivity to negatively framed messages (McClure et al., 2009).

Related research has shown that scenarios that refer to structural features that increase earthquake damage influence people's attributions for the damage (McClure, Sutton, & Sibley, 2007). These authors presented two versions of a news report, one accurate version based on reports by engineers, and one inaccurate version reflecting fatalistic news reports, but presented as being from an engineer. The accurate report read: "The chief city engineer stated that most buildings that collapsed did not meet the current building regulations relating to withstanding earthquakes. These buildings were made of brick or concrete without steel reinforcing. This type of construction is brittle and vulnerable to earthquakes." In contrast, the inaccurate report read: The chief city engineer stated that "several buildings that collapsed met the current building regulations relating to withstanding earthquakes. They were modern buildings made of contemporary building materials. This type of construction is meant to withstand earthquakes." People who read the accurate message attributed the damage to building design more than those who read the inaccurate message. These findings show how media messages about the specific mechanisms that mitigate damage affect people's judgments about the damage.

McClure, Sutton, and Sibley (2007) extended this research on mechanism and design information by comparing the effects of two types of messages. McClure and colleagues presented two versions of a news report. The first was a fatalistic message describing instances where well-designed buildings were damaged, which is the type of message that is often presented in the mass media immediately after a disaster such as the Kobe earthquake (Cowan et al., 2002). People's judgments are often unduly influenced by these instance-based messages which focus on a single event or outcome (Baron, 2000).

The second type of message was designed to reflect accurate rate-based messages stating that a higher proportion of well-designed buildings were resilient, which is the type of message often communicated by feature articles at some time after a disaster, when engineers and other authorities have analyzed the patterns of damage based on reports by engineers. To balance the two conditions, the inaccurate instance-based version was presented as being from an engineer. The accurate rate-based report read: "The chief city engineer stated that most buildings that collapsed did not meet the current building regulations relating to withstanding earthquakes. These buildings were made of brick or concrete without steel reinforcing. This type of construction is brittle and vulnerable to earthquakes." In contrast, the inaccurate instance-based report read: The chief city engineer stated that several buildings that collapsed met the current building regulations relating to withstanding earthquakes. They were modern buildings made of contemporary building materials. This type of construction is meant to withstand earthquakes." Participants made less fatalistic attributions and inferences after reading the rate-based accurate message than with the instance-based inaccurate message. This effect was found regardless of whether the source of the message was a reporter or an engineer.

It is clear from the preceding text that the information civic and societal (e.g., risk management, media) agencies make available represent significant sources of the raw material citizens draw upon to construct their representations of and beliefs about hazards and risk. The process of interpretation does not, however, stop here. Information from civic and media sources does have a bearing on people's beliefs, but not necessarily in a prescriptive way.

Some people will accept civic and media interpretations at face value. Others may disagree with it and interpret it for themselves.

Once (already processed and filtered) information enters the public domain, it is subject to additional interpretation as it interacts with people's pre-existing mental models, beliefs, goals, and expectation (see Chapter 2). When this happens, the social construction of beliefs at a community level can result in people developing opinions that can help or hinder their ability to deal with threat posed by natural hazard events.

This issue is picked up in the next section. It commences with a brief overview of work that illustrates how people's discussions of hazard issues with others in their community influences whether they decide to prepare or not. It then goes on to discuss how changes in the contemporary social contexts in which people find themselves are changing, and changing in ways that may be reducing opportunities for people to discuss and exchange hazard information. Understanding these changing dynamics has important implications for risk management. This discussion further reiterates the need to engage with community members and identifies key issues that need to be accommodated in risk management programs in order to do so.

However, before embarking on this discussion of changes in contemporary community dynamics, the text returns to the fact that at any point in time, people will have been exposed to the agency and media information discussed above. As this discussion pointed out, exposure to media coverage can change the way people think about and interpret hazards and what can be done to manage their risk. The next section opens with a discussion of studies that explore this from a community member perspective.

COMMUNITY CONTEXTS

The fact that people have limited, if any, experience of hazard events (especially at the higher end of the intensity spectrum) makes it more likely that the contents of people's discussions of hazard issues reflects their interpretation of agency information and media coverage. This is the inferred starting point for the discussion in this section.

For example, in a longitudinal study of earthquake preparedness, Paton, Smith, and Johnston (2005) discussed how one variable, critical awareness (which encompasses the degree to which people talk with others about hazards), was the best predictor of both deciding to prepare and deciding not to prepare (see also Chapter 3). It was inferred from this that some people were acting on information in a more objective manner (i.e., they were taking steps to prepare). Others, however, were not doing so. The fact that some people were not preparing could have been the result of their developing beliefs regarding the futility of preparing from their exposure to negative media coverage (see above). Furthermore, their subsequent discussions with others in their community tend to amplify and confirm these beliefs and legitimize their not taking steps to prepare (Paton et al., 2005). However, this is not the only possible explanation.

It is possible that people's decisions not to prepare could also emanate from the content or subject of discussions with those with whom they interact with on a regular basis. It was not possible to explore this issue in the Paton et al. (2005) study. However, finding a similar relationship in a subsequent study of wildfire preparedness (Paton, Kelly, Bürgelt & Doherty, 2006) provided an opportunity to explore how people's beliefs about preparing developed. The interviews they had with residents in high-fire-risk areas provided support for the contention that people could be characterized as falling into "preparing" or "not preparing" groups, with the members of each group thinking about and discussing wildfire risk and preparedness in different ways.

Paton et al. (2006) noted that preparing was associated with the degree to which people engaged in community life. In particular, participation in day-to-day activities (e.g., neighbors sharing memories of wildfire events when meeting on the street or when involved in community activities) were identified as providing opportunities for people to become more knowledgeable about the wildfire history of the area and to better understand their risk. Decisions to prepare were also influenced by people having a sense of shared responsibility for community safety (a finding which reflects people's degree of connectedness to others—see discussion of sense of community below). This work illustrates how actively participating in community life provided a social context in which people could discuss why and how to prepare. In contrast, the social experiences of those who had decided not to prepare were quite different.

For those who chose not to prepare, reasons for reaching this decision included factors such as disagreement amongst family members regarding the need for or benefit of preparing, conflict with neighbors regarding actions that could adversely affect their environment (e.g., reducing fuel loads had negative connotations in that they destroyed trees), and discussions with others that fostered the belief that preparation was unnecessary or less important than other issues facing the community. Those disinclined to prepare also discussed how the negative opinions of others (e.g., friends who did not believe that preparing was effective, or who felt that environmental protection should prevail over controlled burning) influenced their deciding not to prepare. Thus it is possible that people's social exchanges and interactions can lead to the development of beliefs and practices that can override civic pronouncements regarding preparedness.

These examples illustrate a need for risk communication and public education strategies to complement the provision of accurate and relevant hazard and risk information with the development of opportunities for the level of discussion required to effectively utilize "community" as a resource for facilitating sustained preparedness (see also Chapter 3). However, it cannot be taken for granted that appropriate social contexts in which people can develop shared understanding exist. To appreciate why, it is necessary to understand how social dynamics and characteristics influence are changing the social contexts in which people interact.

Given the important role social interaction and social exchange plays in facilitating sustained preparedness, it is important to understand the social dynamics that operate in contemporary communities and that can affect the availability of "community" as a risk management resource and how any deficits in this regard can be remedied through risk management strategies that seek to engage people in risk management planning and intervention. In contemporary urban environments, several dynamics are changing the characteristics of social life. Some that are relevant for understanding risk communication are discussed here. These are the commons dilemma, changes in opportunities for developing sense of community, and shifts in the nature of community in contemporary society.

The Commons Dilemma

The effectiveness of risk management often entails directing public funds to mitigation (see Chapter 2) and requires everybody to prepare to the same extent. The former raises issues about equity because some people profit more than others from mitigation (e.g., flood mitigation benefits those closer the river). The latter raises equity issues if people believe others are not preparing and so not bearing the same costs (Frandsen, Paton, Sarkassian, & Killalea, 2012). People's interpretation of whether common resources are used in fair and equitable ways and whether they believe other people are acting in equitable ways has a significant bearing on their actions. This introduces social exchange processes as important issues in risk management planning (Bishop et al., 2000).

People's judgments about what is fair and equitable are not made objectively. For example, attributional and social cognitive (e.g., unrealistic optimism) processes introduce bias into people's estimates of fairness, how they consequently estimate the costs and benefits of mitigation actions (see Chapter 2), and the degree to which they support them. It is possible to speculate that if people are being asked to incur costs to prepare, their decision to do so could be influenced by whether others are making similar sacrifices or they could become unwilling to act unless they believe others are taking similar actions. If they do not believe this is so, they may not prepare. This kind of thinking was found in a study of wildfire preparedness (Frandsen, Paton, Sarkassian, & Killalea, 2012). The need to consider this social exchange process in risk management planning is evident in finding that people's support for activities requiring collective action can also be affected by the so-called commons dilemma (or the tragedy of the commons) and can be illustrated using climate change (Kok & de Coninck, 2007).

The commons dilemma refers to social dilemmas where citizens share a common resource and where they gain individually if they take more than their share of a resource and thus have an incentive to do so (i.e., to defect from seeking common good). When one person does this, their individual actions make little difference to the overall outcome, but if many do so, the common resource is destroyed. Thus in a commons dilemma, everyone is better off if all the individuals in the community cooperate and do not defect, yet every individual has incentives to cheat or defect. The term "commons dilemma" derives from the mediaeval commons where people could each graze one cow on the commons; there is an incentive for each individual to covertly graze a second cow, but if many or all individuals do this, the commons is destroyed (Hardin, 1968). Thus individual and communal motives often conflict.

The commons dilemma is played out vividly with climate change, because the earth's climate is a globally shared resource (Lorenzoni, Nicholson-Cole, & Whitmarsh, 2007; Milinski, Semmann, Krambeck, & Marotzke, 2006; Pfeiffer & Novak, 2006; Stoll-Kleemann, O'Riordan, & Jaeger, 2001). At the individual level, many people are reluctant to make sacrifices in their comfortable lifestyle when they perceive that the responsibility for climate change is not being shared by other citizens (Lorenzoni et al., 2007). Stoll-Kleeman et al. (2001) found that participants reported strong agreement for statements like: "I alone can do nothing. I can achieve something only if the others join in." Climate change is also an instance of an environmental issue that involves a social dilemma with a temporal dimension: specifically, the conflict between the individual's short-term interests and the community's long-term interests (Milfont & Gouveia, 2006).

The commons dilemma reflects an implicit social contract where people are unwilling to take actions unless others take equivalent actions. People think that their own actions or their country's actions will make little or no difference to climate change unless other persons or nations also take similar actions, particularly if other persons or nations are much larger or are greater emitters than themselves. This is an example of how people in Western countries apply the equity principle to social exchange processes. In this case, it reflects people believing that large polluters should make greater contributions to the solution. If this is not perceived to be the case, people seek to redress the balance by, for example, doing less themselves.

One recent study analyzed people's perceptions about climate change and their rating of different reasons for not acting in regard to climate change (Aitken, Chapman, & McClure, 2011). A prominent reason for not acting was their feeling powerless to make a difference. Another reason took the form of the commons dilemma and reflected people's unwillingness to take action when others were not taking similar action. Other reasons for inaction included uncertainty about how their actions could make a difference. The study also examined participants' perception of the urgency of the issue of climate change and their view as to whether climate change was a natural phenomenon or anthropogenic (caused by human actions).

People's judgment of the risk from climate change was closely related to their judgment that climate change is anthropogenic, and these two variables together were the strongest predictors of whether participants had taken action in relation to climate change. The commons dilemma and powerlessness were significantly correlated, indicating that for climate change, people's feelings of powerlessness are intertwined with an unwillingness to make changes and sacrifices unless others do the same.

Powerlessness was implicated as a significant constraint on whether people had taken actions in relation to climate change. This finding suggests that people not only feel powerless to make a difference to a global problem, but they are also affected by the issues of fairness in contributions, and whether others are willing to take actions, a distinct but related issue. Indeed, the commons dilemma was a stronger predictor of whether people had taken action in relation to climate change than powerlessness; and when the commons dilemma was included in the regression analyses predicting action, powerlessness was no longer a significant predictor of action. This finding suggests that people are unwilling to take actions in regard to climate change if others do not do so and incur a similar level of sacrifice, cost, or inconvenience. As outlined at the start of this section, this has implications for all aspects of risk management.

If people do not perceive equity in how the costs and benefits of mitigation and preparedness are distributed, they may alter their behavior. This can involve getting others to do more or, more usually, their doing less (Frandsen et al., 2012). The problem is that perceived unfairness is more likely to be the norm (e.g., as a result of the operation of attributional and other cognitive biases), particularly if people are not engaging with others at a level that allows fairness issues to be raised and dealt with (e.g., if people do not engage with others regularly to discuss hazard issues, they will not know what others are doing). Thus, for equity issues to be recognized and resolved and collective action sustained, people need to be actively engaged with others within the risk management process. This, in turn, is more likely to occur in contexts in which people share common interests. That is, collective action to deal with natural hazard risk is more likely to occur when people feel that they are part of a community (Becker et al., in press; McIvor et al., 2009). Community is an important construct in a risk management environment in which social constructions of risk and preparedness play significant roles in determining both how people think about hazards and what they do to manage the risk posed by them. The more opportunities people have to engage with like-minded people (i.e., when they are members of a community), the more likely they are to have meaningful exchanges about hazard issues. The next section explores several ways in which community life can help or hinder the development of people's risk beliefs. If "community" is to realize its potential as a crucible for change and a context for facilitating sustainable actions, it must first be defined.

COMMUNITY

Sarason (1974) defined community as "a readily available, mutually supportive network of relationships on which one could depend" (p. 1). Other definitions describe community in terms of its possessing a network of relationships and interdependencies involving people who share experiences through regular interaction (Tönnies, Harris, & Hollis, 2001). In espousing qualities such as "supportive" and "interdependence," these definitions of community illustrate its importance as a context in which the social goals of risk management can be realized. That is, it offers a context in which goals such as promoting shared responsibility for adopting complementary roles in risk management are more likely to be realized.

Sarason also identified the role of perceived similarity with others as having an important bearing on the nature of community. This underpins the importance of community as a context in which shared social representations of risk and related constructs can develop (whether accurate or not) and in which people are more likely to be motivated to work together to face challenges in life. When this happens, it results in the development of Sense of Community.

Sense of Community

Sense of community encompasses members' feelings of belonging, the belief that members matter to one another and to the group, and the existence of a shared faith that members' needs will be met through their commitment to be together (McMillan & Chavis, 1986). A substantial body of research attests to how social network membership, social cohesion, and social capital contribute to an overall sense of community (Alesina & La Ferrara, 2000; Flora, 1998; Forrest & Kearns, 2001; Morrison, 2003; Portes, 1998; Putnam, 2000).

Sense of community has been implicated in the development and application of preparedness measures and to how people respond to hazard events (Paton, Millar, & Johnston, 2001; Rogers, 1995). When sense of community is high, others within that community are more likely to be regarded as credible and trusted sources of information (Lasker, 2004; McGee & Russell, 2003). When these others have local hazard experience and knowledge, they become an important resource for developing shared representations of risk and its management and for facilitating the development and maintenance of locally-relevant risk management activities (Frandsen et al., 2012). Knowledgeable residents help others identify what to do to manage their risk, and act as a contact and liaison point linking the community and risk management agencies (Indian, 2008; Paton, Buergelt, & Prior, 2008). In practice, however, the degree to which the benefits of a sense of community can be realized is complicated by certain characteristics of contemporary societies. These problems and their implications can be illustrated by reference to how changes in occasions for social interaction in contemporary cities, increasing mobility and community diversity, and the gradual shift in conceptualizing community from a locational to a relational entity affect opportunities for both the maintenance of hazard knowledge in a locality and its social dissemination in ways that serve to sustain hazard preparedness.

Opportunities for Social Interaction

The quality of social interaction can be affected by more than just relationships between people. It can be influenced by the characteristics of people's physical environment; particularly the local availability of common spaces in which people can meet informally. Factors such as population growth and development pressures are contributing to a dwindling in the availability and use of common spaces (e.g., neighborhood parks) in many contemporary cities. On the face of it, this may appear to have little to do with risk management. However, it does, albeit indirectly.

The loss of common spaces decreases opportunities for regular interaction within neighborhoods and, consequently, reduced prospects for developing and sustaining sense of community and the informal sharing of hazard knowledge (Prior, 2010). This notion is supported by the work of Kuo and colleagues (e.g., Coley, Kuo, & Sullivan, 1997; Kuo, Sullivan, Coley, & Brunson, 1998; Kuo & Sullivan, 2001).

Kuo and colleagues identified how shared neighborhood spaces were crucial for developing a sense of community and for facilitating interaction and discussion in ways that increased people's capacity to deal with challenging events in their lives. While not relating to natural hazards *per se,* the work of Kuo and colleagues could be equally applicable to issues of hazard preparedness.

Shared neighborhood meeting places foster community participation (see Chapter 7) and opportunities for building and sustaining personal relationships and for sharing information in ways that help build understanding through regular and meaningful engagement with like-minded others. If informal meeting places are lost, so too are (regular) opportunities for sharing stories. Changes in the availability of common spaces are not the only factors capable of affecting how relationships within community contexts are diluting opportunities to maintain local hazard knowledge and preparedness. Others include the increasing mobility of urban populations and increasing population diversity.

Mobility and Diversity

Migration and turnover of residents within an area contribute to increasing diversity in people's environmental beliefs, including those about hazards (Brenkert-Smith, Champ, & Flores, 2006; Carroll et al., 2005; Cottrell et al., 2008; Grothmann & Reusswig, 2006; Hodgson, 2007; Keller, Siegrist, & Gutscher, 2006; Sjöberg, 1979, 2007; Tierney, 1999). As diversity increases, so does the challenge to risk communication. As social representational diversity increases, and local hazard beliefs are diluted, the more heterogeneous become the interpretative frameworks that will prevail within a neighborhood or geographical area and the more challenging the risk communication environment becomes (see Chapter 2). The impact of the social changes created by resident turnover and becomes more apparent following an examination of the more tangible aspects of local life such as those associated with housing tenure.

DiPasquale and Glaeser (1999) showed that home ownership, and particularly the length of residence in a home (and thus in a geographical location), predicts the development of social capital in a community. Not surprisingly, home ownership and length of residence shows stronger links with hazard experience, with both being associated with greater levels of hazard preparation (Spittal, McClure, Walkey, & Siegert, 2008; Tanaka, 2005). Hazard experience, information sharing between long-term members of the community, and heightened responsibility of protecting the investment in the family home interact to increase the likelihood that people will prepare compared with neighbors who rent their homes. Homeowners develop stronger relationships with their neighbors (DiPasquale & Glaeser, 1999; Jakes, 2002; Lichterman, 2000; Low & Altman, 1992; McGee & Russell, 2003; Paton, Buergelt, & Prior, 2008) and share hazard preparation information more freely with other long-term neighbors they know well, thereby increasing the likelihood that community members in these social networks will prepare (McGee & Russell, 2003; Paton, Buergelt & Prior, 2008).

Consistent with the views expressed by Kuo and colleagues (see above), Alesina and La Ferrara (2000) discuss how sense of community is facilitated by participation in community activities (which, for example, increases trust and social capital) but is degraded by community heterogeneity (Ley & Murphy, 2001; Trewin, 2006). Alesina and La Ferrara describe how in communities of place (see discussion of locational communities below), an inverse relationship between levels of heterogeneity (in culture, age, experience, etc.) and active participation in activities with other community members is often found. A reduction in local hazard knowledge and the lack of availability or the dilution of social networks to sustain and communicate this information can make it more difficult to sustain a culture of hazard preparedness in a given area.

Morrison (2003) concurs and discusses how sense of community is diminished by factors such as declining housing tenure, resident turnover, and growing diversity in cultural identity. High turnover of residents tends to diminish social cohesion and sense of community, erodes familiarity and trust (see below), and increases levels of community fragmentation (Morrison, 2003). High resident turnover tends to degrade the membership and quality of social networks (Forrest & Kearns, 2001; Morrison, 2003). The social fragmentation that accompanies turnover increases social exclusion which, in turn, reduces levels of individual commitment to the location or neighborhood where people live, increase community vulnerability in the event of a hazard, and limit opportunities for the kind of sustained social interaction required to develop "hazard mental models" that support mitigation initiatives and preparedness actions (Morrison, 2003; Paton & Bishop, 1996).

As people's identification with relational communities increases, the perceived salience of issues (e.g., about hazards) within neighborhoods, compared with communities defined by location, decreases (Graffy & Booth, 2008). The development of relational links can be fuelled by turnover and increasing diversity. Alesina and La Ferrera (2000) point out that as community diversity increases, levels of participation in (neighborhood) community groups decline and people increasingly search for opportunities for affiliation with those outside their neighborhood. This results in fewer active social networks within a locality and increases people's reliance on social networks outside of their suburbs (Forrest & Kearns, 2001; Morrison, 2003). A consequence of this is reduced access to local knowledge and experience (e.g., regarding hazards). While people still have access to socially rewarding relationships, these relationships may be less capable of promoting the development of the hazard beliefs required to sustain a culture of preparedness in localities (which is where hazard events will occur) susceptible to experiencing infrequent hazard events (Indian, 2008; Paton et al., 2008). This shift in the locus of social networks has important implications for understanding the nature of community in contemporary urban environments and for understanding hazard preparedness.

Given that most hazards tend to impact on specific geographical locations, *community,* from a risk management perspective, tends to be seen as locational. However, while people may live side-by-side in locations susceptible to experiencing hazard events, this does not mean that they identify with their neighbors at the level required to constitute a coherent source of information or sustain collective action to manage risk. Furthermore, as alluded to above, changes in the availability of opportunities to develop relationships in which hazard knowledge is shared within neighborhoods can also arise from changes in how community itself is construed.

Locational and Relational Communities

For some people, community is synonymous with neighborhood. That is, they enjoy a network of relationships and interdependencies with other people within the immediate geographical area in which they live. This is particularly so for those living in rural areas. For others, community is a more diffuse concept and one defined by social relationships organized around similarities of belief, culture, or interests with people who may be spread over a large geographical area (Bell & Newby, 1971; Elias & Scotson, 1974). These different perspectives introduce the fact that it is possible to differentiate locational (communities of *place* or *territory*) from relational (communities of *interest*) communities (Gusfield, 1975).

Relational communities comprise people who come together of their own volition and do so because they share similar values, goals, or interests.

Thus, for members of relational communities, their sense of identification (community) with others derives from shared interests rather than from where they live (though they can coincide). This sense of identification supports the development of shared risk beliefs and provides them with access to a network that can disseminate information to its members in ways that are more likely to be consistent with people's needs and expectations (Bell & Newby, 1971; Putnam, 2000; Rogers, 1995; Tönnies, Harris, & Hollis, 2001). While membership of a locational community is equally valuable to its members, membership is typically more prescribed (Flora, 1998; Völker, Flap, & Lindenberg, 2007).

A construct that may overlap with the notion of locational community is place attachment. A relationship between levels of interaction with neighbors and emotional attachment to the community where people live has been found (Forrest & Kearns, 2001; Morrison, 2003). Both place attachment and sense of community have been found to play roles in the social construction of risk and levels of community hazard preparedness (Dake, 1992; Frandsen et al., 2012; Hannigan, 2006; Holstein & Miller, 2006; Lupton, 1999; Lupton & Tulloch, 2002; Paton, Buergelt, & Prior, 2008; Tierney, 1999). Place attachment played a significant role in how residents in Christchurch, New Zealand, adapted to the consequences of the 2011 earthquake (Paton, 2012). Residents with a strong sense of place attachment were better placed to develop social networks capable of supporting their recovery. In contrast, residents with fewer locational commitments (e.g., some inner-city residents, students) were less well placed in this regard.

Hummon (1992) and Low and Altman (1992) argue that place attachment (the degree to which people feel that they are embedded within their physical environment) increases people's emotional investment in their community. This sense of emotional investment could provide people with an impetus to prepare for hazards in order to protect salient facets of a place one values (Paton, Kelly, Buergelt, & Doherty, 2006). If community members share feelings of geographic attachment, then they may be more likely to develop social networks that see them participating in communal activities, which might extend to sharing information about hazard preparation, and assisting and supporting each other in protective activities (Forrest & Kearns, 2001; Paton, Buergelt, & Prior, 2008).

Risk Management Issues

It is important to accommodate the locational-relational distinction in risk management for several reasons. These include the fact that the past few decades have witnessed a shift from community being predominantly locational to being predominantly relational (Forrest & Kearns, 2001; Trewin, 2006). For example, in Auckland (New Zealand), some 68% of people describe their community as relational. This reflects the influence of societal changes such as increased personal mobility (see above), changes in career aspirations, and communication technologies that allow the maintenance of social networks even when members are geographically dispersed (Forrest & Kearns, 2001; Morrison, 2003; Putnam, 2000; Trewin, 2006).

A second reason for accommodating locational-relational issues in risk management can be found in the role that cohesive social relationships play in forging risk beliefs and developing local risk management initiatives (see Chapter 7). Another reason derives from the fact that hazards are locational phenomena. If the qualities of social networks required to facilitate people's risk management responsibilities are being diluted by a progressive shift towards relational communities then community engagement strategies need to compensate for the consequent loss of locational risk management resources. One way of achieving this is to add a locational identity for those whose sense of community may be predominantly relational. The prospect of facilitating such shared community identities is aided by the fact that they are not mutually exclusive.

Integrating Locational and Relational Identities

It is possible to engage community members in ways that support their developing a superordinate "hazard identity" defined in terms of the common (hazard) fate shared by those living in specific areas within risk communication and community outreach programs (Flora, 1998; Portes, 1998; Prewitt Diaz & Dayal, 2008). That is, to develop some sense of locational identity in those whose community is predominantly relational. This sense of common fate can enhance the likelihood that residents in an area learn about risk and preparedness issues from others (Paton, Buergelt, & Prior, 2008). Social networks and relationships which develop in response to hazard threat can facilitate the development of the collective recognition of threat and encourage collective action to manage risk. Discussions can then be encouraged within these emergent social networks in ways that increase the likelihood of hazard issues being incorporated into the social and cultural fabric of the community. When this happens, a sense of social responsibility for preparedness (i.e., interdependencies between people extends to public safety) is increased and people's new-found, socially constructed knowledge is more likely to be incorporated into mitigation action (Berkman, 1995; Clarke & Short, 1993; Cottrell et al., 2008; Kohler, Behrman, & Watkins, 2007; McIvor et al., 2009).

A sense of locational interest in hazard preparedness may be strengthened if activities are designed with the involvement of respected community members who had good understanding of both local issues and the strengths and needs of local residents (Dalton, Elias, & Wandersman, 2007; Lasker, 2004), with the credibility of these community leaders deriving from how they used their local knowledge and their social networking and management skills to reconcile mitigation and preparedness options with people's needs and concerns. This view has been supported by work on preparedness for earthquakes (Paton et al., 2005), wildfire (McGee & Russell, 2003), and tsunami (Johnston et al., 2005). The development of community leadership within a preparedness strategy can produce benefits that extend beyond it having a role in preparedness. Community-based leaders with good local knowledge and social connections were identified by Christchurch residents as a particularly valuable recovery resource, and one that made a substantial contribution to resilience following the 2011 earthquake (Mamula-Seadon, Selway, & Paton, 2012; Paton, 2012).

The above discussion identified how contemporary social and societal dynamics can dilute the utility of "community" as a resource for risk management. In particular, it introduced the need to include in risk management strategies that engage people (e.g., around the development of a hazard identity). That is, risk management activities need to encourage people to engage with others in meaningful ways and to encourage people and communities to engage with civic sources of information and resources in ways that are meaningful for community members (Frandsen et al., 2012). Engagement-based strategies can reduce problems associated with top-down and media-filtered communications (e.g., media-fuelled fatalism) and contribute to developing comprehensive risk management strategies in which all stakeholders play complementary roles.

COMMUNITY ENGAGEMENT

Relationships between civic agencies and citizens and between neighborhood and community members play significant roles in the development and maintenance of the risk beliefs required to sustain a culture of preparedness. Functional social interaction between stakeholders is the medium through which hazard and risk information are converted into meaningful community knowledge and action. It is, however, hard to achieve this outcome if risk management is conceptualized and implemented in a top-down manner. The latter makes risk management more susceptible to acting in ways that serve agency and societal imperatives rather that the diverse needs of citizens. To develop more inclusive approaches to comprehensive social risk management, it is important that risk management strategies not only make relevant information available but also facilitate ways in which all stakeholders can engage with each other (directly and indirectly) within the wider risk management context. Strategies that seek to create community contexts conducive to mobilizing people's responsibility for managing their risk and at the same create information (exchange) environments in which agencies and the public actively engage with each other (mixing top-down and bottom-up) can rectify the kinds of problems discussed earlier in this chapter (and see also Chapters 2 and 5).

If risk communication is to advance people's understanding of hazards and their implications, ensure that people are adequately informed (within the limits of available knowledge) of relevant issues, ensure risk management processes accommodate diverse needs, and mobilize community action to achieve risk management goals, it must encompass the interactive exchange of information among individuals, groups, and institutions (U.S. National Research Council, 1989). This entails more than just organizing some public discussion or having public officials address public meetings.

Risk communication should facilitate the development of interactive dialogue about risk and its management in a way that ensures that all stakeholders engage with one another and work collectively and in complementary ways (i.e., involving all stakeholders in ways that accommodate their respective roles and contributions) to identify risks and make decisions about how to mitigate and/or prepare for hazard events and their consequences (Jardine, 2008b, 2008c; Maibach & Holtgrave, 1995; Morgan & Lave, 1990; Petts, 2004). This dialogue should encompass all stages of risk analysis, from risk identification to implementation to evaluation (Jardine, 2008c; Petts, 2004; McComas, 2003; Beierle, 2002; Chess, Salomone, & Hance, 1995).

Pursuing this objective involves two things. One involves developing an engagement process that can be incorporated into risk management planning and intervention. This is the topic addressed in the next section. The second issue concerns how to operationalize engagement-based interventions. This involves identifying the community and agency characteristics and relationships that describe how engagement works in practice. The latter is the subject of Chapter 7.

Engagement Processes and Risk Management

The National Environment Protection Council (NEPC, 1999) outlined a set of principles to guide the development of interactive dialogue and engagement with communities. Pivotal to successful engagement, according to these principles, is managing diversity and being able to reconcile the different needs, views, and expectations that come with this diversity into a coherent risk management plan and engagement process (see above). Accommodating diversity is fundamental to letting community members appreciate that their issues are recognized and acknowledged. This, in turn, lays the foundation for building mutual trust, credibility, and respect; thereby facilitating effective preparedness (see Chapter 7).

The first principle advocates involving all stakeholders (members of constituent communities, businesses, etc.) as legitimate partners in risk management and communication. This underpins eliciting the diverse and often conflicting needs and expectations that can exist within a specific jurisdiction. Engagement involves accepting diversity rather than trying to ignore or defuse the community concerns it raises. Constructive conflict can contribute to more comprehensive and representative risk planning by increasing people's sense of involvement, influence, and control in a complex process (Paton & Bishop, 1996). This contributes to enhancing the quality of risk management and increases the likelihood that the strategies it contains are relevant for a community and that adoption of preparedness will be both greater and more likely to be sustained.

The comprehensive eliciting of views and ideas, however, does not mean that all views are accepted or implemented. The effective utilization of the principle of comprehensive stakeholder engagement involves developing and using mechanisms to elicit and prioritize views and ensure that stakeholders appreciate and accept how and why priorities are determined. Effective engagement is thus built on skills in, for example, encouraging inclusive and active involvement in community forums, using methods such as the Delphi technique to manage diverse and potentially conflicting views, conflict management and planning, and prioritizing.

The second principle, identified by the NEPC, is to plan carefully and to adapt the consultation and risk management processes to meet the needs of different stakeholder groups. The third recommendation is to develop a timeline for the engagement process that is realistic (e.g., for organizing, prioritizing, planning, implementation). This may also involve allowing time to develop key competencies and accommodating new developments or changes. The key is being flexible and responsive to changing community and environmental circumstances. This is unlikely to be a swift process, and long-term commitment to the process by both community members and agencies alike is required.

The fourth principle is to actively listen to community members' specific concerns and devolving responsibility for problem definition and solving them. This helps build trust and credibility in the risk management process. The importance of the latter step derives from the fact that trust and credibility are difficult to obtain and, once lost, are almost impossible to regain (Poortinga & Pidgeon, 2004). Thus, engagement is something that must be developed and sustained over the long term. The fifth principle describes the need to develop and implement mechanisms for facilitating collaboration and coordination between diverse stakeholder groups.

The sixth principle involves providing agency personnel, and particularly those working directly with community members and groups, with training in core competencies (e.g., conflict management, problem solving) required to facilitate the development of community engagement and mentor community members through the process (Frandsen et al., 2012). This may also involve providing tangible assistance (e.g., to travel to meetings, accessing to office facilities) to communities and ensuring that written materials are readable and accessible (especially technical detail). Given that community members will also access information from the media (who are generally more interested in politics and risk and danger rather than safety) and other sources (e.g., internet, social media), engagement plans must manage this issue (see Chapter 6).

The seventh principle advocates a need to communicate clearly and acknowledge the legitimacy of different stakeholder views and opinions. Differences in opinions, beliefs, and ideas that emerge from community diversity can be used to provide the foundation for constructive debate. It is also important to accept that some people will not want to be involved in the process and some will be dissatisfied no matter what the process or the outcome. Some of the problems that can arise here can be circumvented by ensuring the mechanisms for prioritizing (see above) are understood and accepted. Finally, it is important to monitor and evaluate the effectiveness of the risk communication and engagement process and ensure the existence of mechanisms capable of using feedback to evolve risk management outcomes over time.

Collectively, the application of these principles can inform the development of engagement plans. Effective plans will outline the allocation of appropriate resources to consultation and communication efforts to facilitate interactive dialogue and the development of mutual understanding with the community. Plans should also describe how all stakeholders will be engaged, identify the techniques that will be used to achieve this, and outline the time schedule for engagement and action to occur.

Achieving effective consultation and communication with stakeholders relies on selecting methods of communication that will reach the target groups. Recognition of the existence of community diversity means that agencies must be prepared to use different approaches to communicate with different groups within a given geographical area (Frandsen et al., 2012). It is also important to recognize that there will be multiple stakeholders and that different techniques will be required to engage each stakeholder group and to develop an integrated plan that reconciles, as far as possible, the needs of all groups within a jurisdiction. Clearly, compared with information dissemination approaches (e.g., pamphlets distributed to mail boxes), engagement-based approaches to public outreach are costly in terms of the time, commitment, resources, and expertise required to develop and sustain the process. However, the benefits of doing so are substantial, particularly with regard to increasing the likelihood of developing sustained community involvement in risk management.

Risk management strategies based on comprehensive engagement increases opportunities to identify issues and misconceptions, facilitate the development of citizen's mental models in ways that reconciles their content with those of their expert counterparts, and enhance the capacity of all parties to communicate, share, and act on risk information in more consistent and complementary ways. Risk communication based on community engagement principles increases the availability of local knowledge, increases citizen's sense of ownership of decision processes, and provides all stakeholders with opportunities to question and explore risk information and its implications, and to discuss how best to apply information to their particular circumstances (Frandsen et al., 2011; Rogers, 1995; Paton, 2008). Furthermore, because it facilitates stakeholders understanding of the need to share responsibility for the development and subsequent adoption of risk management options developed through consensus, it increases both people's acceptance of decisions about uncertain and threatening events and the likelihood of the sustained adoption of mitigation and preparedness measures (Bishop et al., 2000; Graffy & Booth, 2008; Jardine, 2008a).

As a paradigm, community engagement provides a framework for risk management planning. Translating these plans into intervention strategies requires understanding the competencies and characteristics of social relationships (between community members and between them and civic risk management agencies) necessary to convert engagement plans into action. The next chapter introduces how this can be accomplished through applying the concept of empowerment.

Chapter 7

HAZARD PREPAREDNESS: COMMUNITY ENGAGEMENT AND EMPOWERMENT

INTRODUCTION

Chapter 6 concluded by introducing a need to identify the social competencies and characteristics required to turn engagement plans into action. This challenge is picked up in this chapter. A need to understand the role played by social competencies and characteristics in comprehensive conceptualizations of preparedness was alluded to in earlier chapters. For example, Chapter 3 discussed how the development of certain social preparedness functions (e.g., becoming a member of hazard-related organization, actively participating in community preparedness meetings) identified in several studies required the presence of certain social and relationship skills (e.g., willingness to commit time to working with others, actively involving others in discussions, collective problem solving and planning) if they were to develop. Nor is a need to understand social and relationship competencies an issue restricted only to community members. It also applies to understanding relationships between community members and civic risk management agencies. This chapter discusses how community and community-agency relationships (and their implications for preparedness) can be integrated in a theory of community engagement. The main issues being canvassed in this context are summarized in Figure 7.1. In so doing, this chapter builds on a long history of empirical research that has identified how social characteristics and relationships influence hazard preparedness.

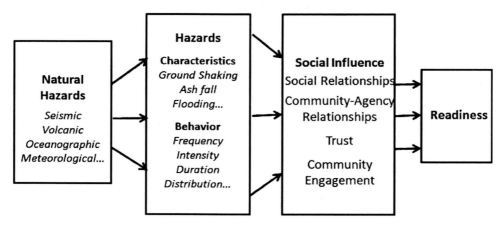

Figure 7.1. Social influences on preparedness.

SOCIAL RELATIONSHIPS AND HAZARD PREPAREDNESS

Social relationships have been implicated in hazard preparedness in numerous studies. Tierney et al. (2001) noted that social cohesion and participation in community activities influenced levels of hazard preparedness. Turner et al. (1986) described how "bondedness" (e.g., length of residence in a neighborhood, identification of the neighborhood as home, participation in community life, and the presence of friends and relatives nearby) predicted preparing for earthquakes. The level of people's involvement in community networks has been identified as a key predictor of earthquake (Heller et al., 2005; Mileti, Fitzpatrick, & Farhar, 1992; Paton, 2003; Tanaka, 2005), volcanic (Dominey-Howes & Minos-Minopoulos, 2004; Gregg, Houghton, Paton, Swanson, & Johnston, 2004; Paton, Johnston, Bebbington, Lai, & Houghton, 2001a; Paton, Smith, Daly, & Johnston, 2008c), tsunami (Johnston, et al., 2005), cyclone (Anderson-Berry, 2003), flooding (Grothmann & Reusswig, 2006; McIvor & Paton, 2007; Siegrist & Gutscher, 2006, 2008), and wildfire (Bright & Manfredo, 1995; McGee & Russell, 2003; Paton, Buergelt, & Prior, 2008a; Prior & Paton, 2008; Vogt et al., 2005) hazards. Indeed, Heller et al. concluded that people's level of active involvement in informal community and neighborhood networks was the strongest predictor of preparedness in a study examining a range of predictors. Similarly, Mileti et al. (1992) and Paton et al. (2005) found that when citizens had been given information encouraging them to prepare, discussion in communities and observing other people's preparations were more important predictors of preparedness than the information distributed. McIvor et al. (2009) and Paton, Buergelt, and

Prior found that having a sense of "responsibility to others" was a significant predictor of preparedness for earthquake and wildfire hazards respectively. These studies provide tangible evidence of the need to include social relationship variables in comprehensive conceptualizations of the preparedness process. What about community-agency relationships?

The importance of including community-agency relationships in comprehensive conceptualizations of social risk management is reinforced when considering how, when faced with uncertainty, and when unable to gain the insights necessary to understand the implications of hazard and their management for themselves, people come to rely on civic agencies (and others) to help them interpret risk information and offer advice and recommendations about how they might manage their risk (Bishop, Paton, Syme, & Nancarrow, 2000; Carroll, Cohn, Seesholtz, & Higgins, 2005; Prewitt Diaz & Dayal, 2008; Vieno, Santinello, Pastore, & Perkins, 2007).

Evidence supporting the benefits of community-agency relationships exists. For example, Frandsen et al. (2012) demonstrated the value of fire agency representatives actively engaging with community members to facilitate wildfire preparedness. Related research has shown the value of scientists gathering with citizens in a participatory process where they discuss and explain their predictions and the uncertainties relating to the science. An example of where this has been successful in relation to climate change has been scientists working with farmers in Africa to make good agricultural decisions (Roncoli, 2006). This strategy not only facilitated interaction between scientists and farmers but also amongst the farmers themselves. In addition, strategies adopted by a minority in a community, can spread to the whole community in appropriate conditions, without requiring the ongoing direct involvement of scientists (e.g., Moscovici, 1968). It is possible to extrapolate from this example to infer the benefits that could accrue from all stakeholders engaging in complementary ways in risk management programs. There thus exists evidence that people benefit from engaging with others in the communities in which they participate in social activities and from engaging on a more equal footing with scientists and civic risk management agencies.

Taken together, the contents of this section clearly indicate that certain characteristics of the relationships between community members and between them and civic (e.g., risk management agencies) sources of information and assistance hold considerable potential as variables capable of informing understanding of preparedness. This chapter builds on earlier discussions (see Chapters 5 and 6) to explore how relationships between people and between people and agencies can be conceptualized in ways that capture their inter-dependent contributions to preparedness. The starting point for this conceptualization was not the community, but rather the nature of the community-agency relationships. It starts by reconsidering how agency roles might be viewed in the context of risk management.

RELATIONSHIPS BETWEEN PEOPLE AND AGENCIES

Historically, the adoption of a top-down approach to risk communication has led to the role of scientists and risk management agencies as purveyors of expert information not being considered other than as sources of information. However, when attention switches to focusing on engagement, a more critical appraisal of the relationship between the recipients of information (community) and the sources of information (agencies) is required. One issue that is highlighted by a critical re-examination of this relationship is a need to consider people's attitudes to and beliefs about the agencies themselves (Paton, 2008).

Some people may accept the information that comes from civic and societal sources at face value. Others will not. For the latter, risk information interacts with pre-existing mental models (see Chapters 2 and 5) that can include elements derived from people's past experience with (e.g., as information sources) or of (e.g., through media portrayals of agency responses to past emergencies or disasters) agencies. This highlights the fact that people have dealings with civic agencies, government departments, and the media, both directly and indirectly, in ways that extend beyond risk management per se and that extend back in time.

The historical experiences people have had with sources of information (e.g., direct experience with government agencies and scientists, indirect experiences with these sources that were filtered through media reporting—see Chapter 6) contribute to the formation of attitudes and beliefs about these sources that develop over time. People sum the experiences with agencies they have had over time to define their relationship in value-laden ways; helpful or unhelpful, empowering or disempowering and so on, with these interpretations informing their beliefs about how much they trust others. Trust is a fundamental aspect of a well-functioning community (Forrest & Kearns, 2001; McMillan & Chavis, 1986; Portes, 1998; Putnam, 2000), particularly when uncertainty increases people's reliance on scientists and risk experts for information and advice.

Trust

Trust is a prominent determinant of the effectiveness of interpersonal relationships, group processes, and societal relationships, particularly when people are faced with the task of dealing with unfamiliar, infrequent, and complex environmental hazards (Kumagai et al., 2004; McGee & Russell, 2003; Paton, 2008; Siegrist & Cvetkovich, 2000; Winter, Vogt, & McCaffrey, 2004). Trust influences people's perception of others' motives, their competence, and the credibility of information provided (Earle, 2004; Kee & Knox, 1970). Consequently, trust plays a pivotal role in defining people's perception of the quality of any relationship in which information is exchanged, appraised, and decisions made, particularly when people are being called upon to make decisions under conditions of uncertainty (Earle & Cvetkovich, 1995; Siegrist & Cvetkovich, 2000).

Levels of people's risk acceptance and their willingness to take responsibility for their own safety are increased, and decisions to take steps to actively manage their risk more likely, if they believe that their relationship with formal agencies is fair and empowering (e.g., agencies are perceived as trustworthy, as acting in the interest of community members) (Bočkarjova et al., 2009; Bishop et al., 2000; Foddy & Dawes, 2008; Lion et al., 2002; Paton & Bishop, 1996; Poortinga & Pidgeon, 2004). When the relationship between people and agency is not perceived as fair, the consequence can be the development of mistrust in agency sources of information (Eilam & Suleiman, 2004; MacGregor, Slovic, Mason, & Detweiler, 1994).

The importance of including trust in conceptualizations of risk management can also be traced to the fact that the greater the degree of novelty and ambiguity present, and the less structured the risk context, the more people attribute weight to their general trust beliefs about, and past trust experiences with, the source(s) of information they turn to or have to rely on (Johnson-George & Swap, 1982; Luhmann, 1979; Prior & Paton, 2008; Siegrist & Cvetkovich, 2000; Sjöberg, 1999; Worchel, 1979). This means that the quality of trust in agencies (and others) developed under normal circumstances (i.e., in relation to everyday dealings with sources such as government agencies, the media, etc.) become important determinants of trust in the context of risk communication about or discussion of infrequent hazard events. The quality of the relationship people perceive themselves as having with a source reflects the balance of positive and negative experiences they have had with that source (e.g., government agency, media) over time (see above). The perceived quality (from community members' perspective) of this relationship will influence people's willingness to use information from civic sources (e.g., risk management agencies, the media) to understand their circumstances, identify what they should do, and evaluate the ability of protective actions to mitigate their risk. What this means is that people's beliefs about the trustworthiness of a *source* (agency) of information can make a contribution to decision making that is independent of the information a source provides.

Taking the above issues together, a comprehensive theory of preparedness needs to encompass how relationships within communities and relationships between people and civic sources of hazard and risk information influence preparedness. This chapter discusses a theory that attempts to do so. This theory is built on two separate but related constructs: empowerment and trust. These constructs come together in the sense that if past experiences resulted in people construing their relationship as being fair and empowering, people are more likely to trust a source and more likely to use the information provided by a source in the manner intended; that is, to develop and sustain people's preparedness. If, on the other hand, people do not trust a source, they are more likely to question the validity of the information or the motivation of the source and, consequently, are considerably less likely to use information from this source for any reason.

The importance of including the empowerment construct in a conceptual framework derives from it comprising two components (Dalton, Elias, & Wandersman, 2001). These are depicted in Figure 7.2. In Figure 7.2, the provocation is held to be the need to address the threat posed by natural hazards. How this is accomplished, according to this conceptualization, is through the action of two separate but complementary processes. One of these, empowered communities, captures the need for communities to possess certain characteristics if they are to be self-determining. The other, empowering settings, describes how agencies and societal institutions can create the circumstances and act towards people and communities in ways that facilitate the ability of community members to take responsibility for managing the challenges encountered in life.

This conceptualization depicts the goal of community engagement as creating empowered people and empowering settings. These play complementary roles in the risk management process; both are required if engagement strategies are to be successful and sustainable. When this happens, the relationship between people and agency sources is more likely to be one characterized by trust. In order to turn this conceptualization into a reality, it is necessary to identify the characteristics of empowered people and those of empowering social settings and how they interact through the medium of risk communication and community outreach initiatives to facilitate community preparedness. How this can be accomplished is illustrated in this chapter in the form of a community engagement theory (Paton, 2008).

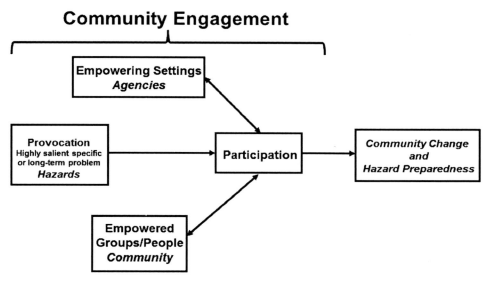

Figure 7.2. The relationship between empowered people and empowering settings (adapted from Dalton et al. (2007)).

THE COMMUNITY ENGAGEMENT THEORY

This chapter illustrates how empowerment can be operationalized using a community engagement theory that depicts trust as a process influenced by dispositional (e.g., expectations of outcomes), situational (familiarity, information availability), and structural (e.g., levels of participation in community life) factors (Paton, 2007; 2008). The conceptual foundation for this theory is illustrated in Figure 7.3. The inclusion of situational and structural factors (e.g., familiarity, information, participation levels) provides a basis for exploring how interaction between community characteristics and trust defines the context within which information about hazards is appraised and decisions regarding preparedness made. In the next section, the function and roles of the variables are described. This process commences with a discussion of how personal beliefs (outcome expectancies) interact with social processes (community participation) and competencies (collective efficacy) to contribute to the development of empowered communities.

Empowered Communities

The theory proposes that preparedness decisions are founded on people's beliefs regarding the perceived effectiveness of the preparedness mea-

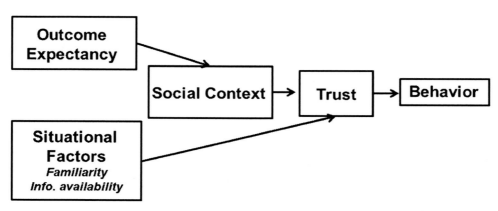

Figure 7.3. The basic trust process (Adapted from Kee & Knox, 1970; Mayer et al., 1995; Paton, 2008).

sures described in public education programs (see also Chapter 6). Public education programs essentially advise people that if they adopt a particular behavior the *outcome* will be a reduction of their risk or an increase in their safety. However, in the uncertain environment of hazard preparedness, people interpret this information (e.g., they cannot readily test the veracity of the information for themselves) and its recommendations (e.g., beliefs about the magnitude of events and what this could mean for the controllability of hazard events, exposure to media coverage that has typically emphasized how catastrophic hazard consequences are, etc.) to estimate whether they believe or *expect* that *outcome* (i.e., increased safety, reduced damage, etc.) to occur. The "outcome expectancy" construct describes this interpretive process (Bennet & Murphy, 1997) and it can be further subdivided into negative and positive outcome expectancy. Negative outcome expectancy has effects that are comparable to those described for fatalism (see Chapter 4). However, negative outcome expectancy relates to people's beliefs about a hazard and its consequences rather than reflecting a fundamental life view (fatalism) that is applied to all aspects of a person's life.

Negative outcome expectancy describes an interpretive process which culminates in people forming the view that hazard consequences are too catastrophic for personal action to make any difference to peoples' safety. Holding this belief reduces the likelihood that people will prepare (Paton, 2008). In contrast, if people hold positive outcome expectancy beliefs, they are more likely to believe that personal actions can enhance personal safety and/or mitigate hazard consequences. However, a belief that preparing can be effective does not necessarily (i.e., it is an expectancy) equate with having full knowledge about what to do or how to put ideas into practice.

If people need additional guidance to deal with their uncertainty, they look first to other community members. For instance, people may want to first know what their neighbors or others in their community have done, check others' views about what might work, and identify what are the most important measures to adopt and so on. For example, participants in Paton, Buergelt, and Prior's (2008) study of wildfire preparedness reported how day-to-day participation in community life and activities (e.g., neighbors discussing previous wildfires when meeting on the street or when involved in community activities) informed their understanding of the wildfire history of their suburb and helped them work out why and how to prepare. This latter finding identifies how interaction with others can provide both access to sources of information and serve a problem-solving function (e.g., discussion with others can help people to work out what to do and why). This illustrates how the level of people's engagement with social groups and networks influences how their uncertainties are confirmed or resolved (see also Chapters 5 and 6), and their understanding given form in a manner consistent with their (local) needs and expectations. Thus, social interaction influences how risk beliefs are developed and sustained.

Because participating in community activities provides access to information from people who are more likely to share one's interests, values, and expectations, information from other community members can assist in understanding one's circumstances and deciding what to do (Eng & Parker, 1994). This led to including "community participation" as a variable in the community engagement theory.

Community Participation

From a risk management perspective, the importance of community participation can be traced to the fact that when faced with complex and uncertain events, people's perception of risk and how they might mitigate it are influenced by information from others who share their interests and values (Earle, 2004; Jakes et al., 2003; Lion et al., 2002; Marris et al., 1998; McGee & Russell, 2003; Paton & Bishop, 1996; Poortinga & Pidgeon, 2004; Rippl, 2002). People are more likely to trust information when it comes from people with whom they identify and interact with frequently. Interaction and discussion with "like-minded" people is more likely to produce outcomes that are consistent with one's needs and expectations at the time.

The community participation measure used assessed the degree to which people take an active part in community life (e.g., volunteering for community activities, membership of mutual assistance groups, providing support to other community members). The benefits of participation are several and include acquiring new information from discussions with people, learning new skills, being involved with important issues, making interpersonal contacts, personal recognition, and a gaining a sense of improving the community by contributing to improving their own and others' quality of life (Dalton, Elias, & Wandersman, 2007; Earle, 2004).

While participation in activities with other people can provide access to collective knowledge and expertise, deciding how they can take action to deal with unfamiliar circumstances is a function of people's ability to, for example, identify the hazard consequences they could encounter, the specific implications these consequences have for them, and identify what they could do to eliminate or minimize the adverse consequences that hazard activity has for them. This problem-solving capability is essential both for helping people anticipate what they might encounter (within their environment) and helping them identify how they might deal with the risk management issues they have identified. Because the "whole is greater than the sum of its parts," the effectiveness of this process (e.g., the diversity of knowledge, skill, and perspectives brought to bear on the task) is enhanced if performed as a collective activity (Paton, 2008). One construct that captures this competence is collective efficacy.

Collective Efficacy

When dealing with complex and uncertain environmental events, a capacity to formulate questions that can elicit answers that lead to solutions consistent with community members' values, needs, and expectations (e.g., defining community needs, seeking to fill gaps in knowledge or skill) plays an important role in developing people's ability to appraise and evaluate information. Collective efficacy is a construct that provides insights into community members' ability to perform these problem solving tasks.

Collective efficacy describes community members' ability to assess their capabilities and resource needs in relation to challenging tasks, define their goals, and develop plans for using resources to achieve goals when confronting complex, challenging tasks (Bandura, 1997; Duncan et al., 2003; Paton & Tang, 2009; Zaccaro, Blair, Peterson, & Zazanis, 1995). Collective efficacy was thus included in the theory to assess community members' ability to identify the information, resource, and planning needs required to advance their preparedness planning. Because collective efficacy can only develop in established communities, the theory proposes that collective efficacy mediates the relationship between community participation and empowerment.

Community participation and collective efficacy combine to increase the level of resources (e.g., knowledge, skills, etc.) community members have at their disposal and enhance the capacity of community members to use these resources to deal with the challenges they encounter. That is, they make a significant contribution to creating an empowered community. However, when people are faced with atypical challenges, such as those posed by having to anticipate issues arising when dealing with complex and infrequent hazard events, community members may have to acquire information and resources from sources outside their community in order to meet their unique needs and expectations.

A key element in the last sentence was its highlighting the fact that, even when facing the same hazard, communities can differ with regard to their specific risk management needs. For example, in their multi-community study of wildfire preparedness, Frandsen et al. (2012) found that no two of the communities they studied identified the same risk management problem as being the most salient for them. That is, each community had different information, resources, and assistance needs. For each community to advance its hazard preparedness, the agencies engaging with these communities need to be flexible with regard to both how they relate to and how they interact with different communities in order to meet the specific needs of each community.

Recognition of this state of affairs introduces another level of engagement into the theory; one concerned with how agencies relate to communities. The effectiveness of the role of expert sources within a community engagement model is a function of the degree to which agencies create empowering settings (Dalton et al., 2001). It is important that agencies relate to communities in ways that complement community activities.

Linking People and Experts: Empowered Communities Interacting with Empowering Settings

The levels of community participation and collective efficacy prevailing within a community at any given time influence the degree to which its members can engage with one another (whether in locational or relational communities) to formulate, for example, gaps in knowledge and resource needs that must be filled before they can fully comprehend their circumstances and/or be able to act to mitigate or prepare for possible hazard consequences. Once these needs have been identified, people's subsequent ability to act is a function of the degree to which agency and societal institutions empower community members by providing the resources and information that meet people's needs and expectations in ways that reduce people's uncertainty and facilitate their ability to act (Earle, 2004; Eng & Parker, 1994;

Frandsen et al., 2012; Paton & Bishop, 1996; Poortinga & Pidgeon, 2004). In other words, being responsive to community needs is fundamental to the goal of empowering communities in ways that increase the likelihood of their taking responsibility for their safety and adopting their role in a comprehensive risk management plan.

Understanding the relationship between empowered communities and empowering settings is important in other respects. The degree to which empowered communities and empowering settings develop in ways that complement each other influences trust. It is the consistency between the needs and expectations generated by community members (i.e., via community participation and collective efficacy) and the information and resources received from civic sources (i.e., whether the information and resources provided clarify issues identified by community members and/or facilitate action) that creates the empowering settings required to help people construct more accurate estimates of risk, reduce uncertainty, enhance trust, and use information to take action (Earle, 2004; Eng & Parker, 1994; Lion et al., 2004; Paton, 2008; Speer & Peterson, 2000).

Only if people perceive that the information or resources they receive as answering their questions or meeting their needs will it act as a catalyst for action (Paton, 2008). Thus, information could be perceived as empowering if, for example, it fills gaps in knowledge identified by community members or helps develop community competencies (e.g., using field demonstrations to show how to prepare houses for wildfires, working with community leaders to develop competencies in conflict resolution and planning, etc.) that community members identified as being relevant to meeting their needs (Frandsen et al., 2012). That is, community action (i.e., preparedness) is more likely when the information that people receive from agency sources meets a need identified by community members and/or facilitates their ability to respond in a way that is meaningful for them. Eng and Parker (1994) argue that the effectiveness of collective problem solving is a function of the degree to which reciprocal feedback between the parties (e.g., people and agencies) exists to facilitate community members' goal attainment. This is the process encapsulated in the construct of empowerment (see Figure 7.2).

Empowerment

Empowerment describes citizens' capacity to gain mastery over their affairs and confront environmental issues while being supported in this regard by external risk management sources rather than being directed by them or having solutions thrust upon them by agency policy or practice. Empowerment strategies are driven by the goal of promoting the equitable distribution of resources (material, social, knowledge, peer helping, belong-ingness) to facilitate social justice (see discussion of the commons dilemma in Chapter 6), sense of community, and the development of a collective capacity to confront local issues, whether of a hazardous nature or not (Eng & Parker, 1994; Paton & Bishop, 1996).

As introduced above, the degree to which risk management processes (and the strategies they adopt) can be regarded as empowering is a function of two separate but related elements: empowered people and empowering settings (Figure 7.2). In the community engagement theory, factors such as participation and collective efficacy contribute to people being empowered (e.g., because they contribute to people's capacity to identify needs and represent those needs to others). However, if they are to act despite the prevailing uncertainty, actions will depend in part on the degree to which the attitudes and actions of societal agencies create empowering settings. These elements play complementary roles in risk management, especially when people are reliant on sources outside their community (as is the case with natural hazard risk management).

Empowerment thus reflects the quality of reciprocal relationships between community members and between community members and societal institutions and risk agencies. The quality of people's relationship with agencies is defined in part by the degree to which the agency devolves responsibility for planning and action to community members. This, in turn, influences the level of trust that exists between community members and civic emergency planning agencies. That is, the more citizens perceive their needs as having been met through their relationship with civic institutions, the more likely they are to trust them and the information they provide. People's perceptions of the degree to which they are empowered will manifest, at least in part, in the levels of trust they feel toward an agency. The community engagement theory argues that trust mediates the relationship between social context factors and, first, intentions to prepare, then actual preparedness.

In summary, Paton's (2008) community engagement theory proposes that adoption of protective measures is the outcome of a process that commences with people's outcome expectancy beliefs. If people hold negative outcome expectancy beliefs, this reduces the likelihood of adopting protective measures. With regard to factors that predict adoption, it was hypothesized that community participation, collective efficacy, empowerment, and trust mediate the relationship between positive outcome expectancy and intentions, with intentions mediating the relationship between trust and preparedness (Figure 7.4 a-c). The analyses summarized in Figure 7.4 illustrate three examples of tests of this theory for tsunami, earthquake, and wildfire

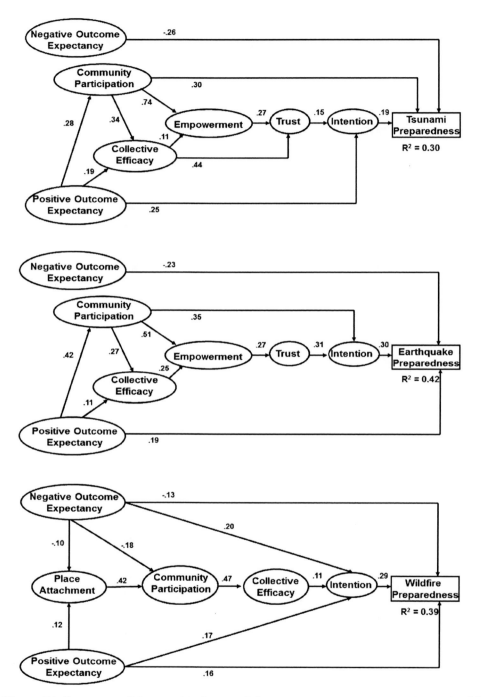

Figure 7.4. Summary of the empirical tests of the community engagement theory for (a) tsunami, (b) earthquake, and (c) wildfire.

hazards (Paton et al., 2009; Paton, in press; Paton, Buergelt, & Prior, 2008).

The analyses summarized in Figure 7.4 provide broad support for this theory being able to contribute to understanding hazard preparedness. Despite some minor differences in path relationships, the variables and the hypothesized relationships between them held across a range of hazards. Analyses accounted for between 30% and 42% of the variance in preparedness. This does not mean that these are the only variables that need to be considered. These analyses demonstrate that social characteristics and processes assist understanding how people relate to their environment (e.g., develop risk beliefs) and how people make preparedness decisions. Taken together, this means that social context factors need to be included in comprehensive conceptualizations of hazard preparedness and in the risk communication and community outreach programs used to facilitate preparedness.

An interesting finding in the wildfire analysis (Figure 7.4c) was that a better fit was obtained if empowerment and trust was excluded. Given the fact that the theory was built around how trust influenced decision making and action under conditions of uncertainty, it is important to discuss possible reasons why a better fit for the wildfire analysis was obtained by excluding empowerment and trust. One possible explanation relates to the fact that the importance of trust in decision making is inversely related to the degree of familiarity people have with a situation (Earle & Cvetkovich, 1995; Luhmann, 1979; Mayer et al., 1995; Siegrist & Cvetkovich, 2000). This is reflected in the inclusion of situational influences in the conceptual model from which the theory was derived (Figure 7.3).

As familiarity increases so does the amount, quality and availability of information relevant to the situation people themselves possess or can more readily access. Support for this view comes from finding that people in the areas from which the wildfire data were obtained were three times more likely to discuss wildfire issues every month compared with their counterparts living in high earthquake risk areas (Paton, 2008). As hazard frequency increases, so does the likelihood of people having direct (e.g., having been affected by wildfire) or indirect (e.g., witnessing fire in close proximity to their community, knowing friends and family who have been affected etc.) experience of the events they are being asked to prepare for. This means that more information will be directly available to people and/or accessible from within their community, negating the need to acquire and evaluate information from other sources.

Note that this does not mean that the information available to people, or that that they selectively attend to, is accurate or capable of motivating action (see Chapter 3). What familiarity means is that people have access to information that influences their interpretation of hazards and the choices they make about managing their risk. Clearly, risk communication needs to accommodate this and ensure that people are basing their choices on accurate information.

With regard to the potential role of familiarity in the area from which the wildfire data were collected, the annual nature of wildfire events in these areas could have increased people's familiarity with wildfires. This reduces reliance on external sources and thus, under this circumstance, it could be postulated that empowerment and trust (the variables that link people and agencies) become less salient as influences on people's preparedness decisions. Dependence on external sources through empowerment and trust is therefore rendered unnecessary when people are able to make direct evaluations of the costs and benefits of preparing based on their personal accumulation of knowledge. This suggestion remains tentative until a more searching investigation of the nature of familiarity and how it can be assessed and tested is undertaken.

In contrast, in the unfamiliar or novel situations, that better describes the situation prevailing in the areas from which the data from the tsunami and earthquakes hazards were obtained; reliance upon external expert sources would be correspondingly greater. Hence the significance of empowerment and trust as predictor variables in the tsunami, earthquake analyses in which people faced uncertainty and were reliant on others to help them deal with this uncertainty.

The wildfire analysis also differed from its tsunami and earthquake counterparts with regard to a better fit being obtained when a measure of place attachment was included in the model. This could reflect the fact that the majority of the population from whom the data to test the model were collected chose to live in close proximity to woodland and bush for lifestyle reasons. This could have made their sense of attachment to this environment an important influence on their preparedness. That the inclusion of place attachment reflects a characteristic of this population, and not a variable which need be included in all locations, is supported by it not being found in a study of wildfire preparedness in Portugal (Paton, Tedim, & Shand, 2012).

The community engagement theory discussed in this chapter demonstrated that social antecedents can be implicated in accounting for differences in levels of preparedness for different hazards (tsunami, earthquake, and wildfire) and in different countries. By offering some all-hazards and all-community applicability, the theory can be used to inform the development and implementation of the risk communication and engagement components of a risk management strategy. This issue is discussed next.

COMMUNITY ENGAGEMENT IMPLICATIONS FOR RISK MANAGEMENT PLANNING

The community engagement theory discussed in this chapter can provide a framework for developing and delivering risk management programs based on community engagement principles. Engagement is about more than just presenting information at public meetings. The theory depicts engagement in terms of the interaction between empowered communities (i.e., communities that possess the competencies to identify and represent their needs) and empowering settings (i.e., agencies facilitate community action by tailoring intervention to meet community needs and working with community members as partners in risk management). This suggests a need to include the assessment of people's capacity to formulate problems and pertinent questions (i.e., the degree members are empowered through internal community processes) in ways that facilitate their ability to direct their information search in meaningful ways and to use information and advice received to advance their own risk management goals (Earle, 2004; Eng & Parker, 1994; Paton, 2008) in risk management programs. The importance of doing so derives from the implications of these competencies being absent.

In the absence of a capacity to formulate questions and thus information needs, people are less able to seek and then evaluate information in ways that act to clarify the uncertainty they face. That is, information received is less likely to be meaningful to people and so less likely to act as a catalyst for action. If information or advice fails to create the expected clarification or direction people are looking for, people tend to attribute the failure to do so to external sources (the actor-observer effect associated with the fundamental attribution error) such as civic risk management agencies rather than to a lack of effort on their part or to the possibility that they did not specify their information needs with sufficient clarity. When this happens, people's trust in that source will diminish as a consequence. Once trust is lost, it is difficult to regain (Poortinga & Pidgeon, 2004).

This discussion reiterates the benefits to risk management that can accrue from seeing it as a reciprocal process (between community members and agencies) and ensure not only the availability of good information but also the presence in communities of the capacities to identify information needs (e.g., information search, problem definition and solving, articulating needs, decision making, and the ability to represent need to agencies) and the ability to use information and resources to meet their needs (Eng & Parker, 1994; Paton, 2008). If these community competencies are absent, a short-term solution can be to use advocates or mentors to facilitate activities such as problem solving and planning (Dalton et al., 2007).

A community engagement approach thus portrays agencies as integral elements in the social context within which people develop and enact their risk beliefs and mitigation options and not just as sources of information. In particular, this means that the agency role accommodates the need to be fair and empowering. What implications does this have for developing risk communication and outreach programs? This question is explored in the next section.

Developing Risk Communication and Public Outreach Programs

Community engagement planning issues were discussed in Chapter 6. The contents of this section complement that earlier discussion by offering recommendations regarding how engagement programs can be implemented in practice. The community engagement theory suggests that strategies must address information content (e.g., outcome expectancy), social context (community participation, problem solving), and people-agency relationship (empowerment, trust) issues. The following discussion follows the order in which variables are placed in the theory (Figure 7.4). Discussion commences with outcome expectancy.

Strategies for Changing Outcome Expectancy

The community engagement theory suggests that risk management strategies should accommodate outcome expectancy beliefs. Risk communication should aim to reduce negative outcome expectancy (NOE) and increase positive outcome expectancy (POE) beliefs. With regard to NOE, the issues identified in Chapters 4 and 6 regarding fatalism and media influence can be used in this context. This includes facilitating people's ability to differentiate between uncontrollable causes and controllable consequences (e.g., emphasize the distinction between the cause of a hazard and the consequences that can be managed). It is particularly important that the news media echo these sentiments. Similarly, strategies that include framing messages in ways that invite people to consider what could be done to assist more vulnerable members of society (e.g., children at school, residents in a home for the elderly) can be effective as a means of countering NOE beliefs. While strategies that focus on countering NOE beliefs can reduce the likelihood of people deciding to do nothing, encouraging people to prepare makes the inclusion of strategies to enhance POE an important facet of risk management and communication programs.

Positive outcome expectancy beliefs can be enhanced by increasing people's hazard knowledge, their understanding of how hazard consequences arise, their appreciation of what can be done to prevent these consequences occurring, and advising how to reduce the adverse impacts of hazard characteristics on people and property (see also Chapter 4). In particular, it is important to explain how each protective action reduces risk and/or contributes to safety (e.g., how securing tall furniture reduces loss and injury from ground shaking, how a defensible space mitigates ember attack, etc.). Including this information facilitates people's capacity to understand what they can do and how and why preparing can be effective. This process can be further assisted by presenting people initially with a small number of items, starting with relatively easily adopted items or actions and introducing progressively more complex activities over time (see Chapter 3) and by embedding the discussion of these issues in social contexts (e.g., community group meetings). By presenting information on preparedness measures progressively over time to social groups and/or encouraging people to discuss issues within the family or with their neighbors, sustained adoption of preparedness measures is more likely (Frandsen et al., 2012). If a strategy is to include activities such as encouraging discussions in social settings, it is important that it ensures that the social competencies to use information effectively are in place.

Adapting to Hazardous Circumstances: Social Network Influences

From the perspective of mobilizing social resources to empower people, a number of challenges to risk communication can be anticipated. For example, risk management programs are not generally geared to actively facilitating community discussion of such things as hazard issues, developing community members' ability to define and resolve their risk management problems, or engaging with (diverse) communities to develop collaborative approaches to confront the threat posed by a natural hazard. These activities thus need to be specifically included in risk communication and community outreach strategies.

In the absence of the requisite community characteristics (e.g., active participation) and competencies (e.g., collective efficacy), the success of outreach endeavors will be significantly muted. To circumvent this, it will be beneficial for risk management agencies to expand their role from focusing predominantly as suppliers of information to include assessing and developing the competencies required to support the achievement of preparedness goals. This does not mean that risk management has to start from scratch in order to develop these processes and competencies.

An alternative approach emerges from appreciation of the fact that the necessary competencies and processes may be present already (e.g., prob-

lem-solving skills developed from planning community activities or participation in social action groups). The variables (e.g., collective efficacy) identified as significant predictors of preparedness in the community engagement
theory (Figure 7.4) were already present in the populations from whom data
were collected. The challenge is then less about developing these competencies from scratch and more about developing strategies to mobilize their
application to hazard and risk management. This raises the possibility that
risk management objectives can be pursued by integrating risk management
strategies with more mainstream community development programs (Anckermann et al., 2005; Paton, 2008; Paton, 2009; Paton & Tedim, 2012; Rich et
al., 1995).

Integrating Risk Management and Community Development

The main reason for proposing that normally disparate programs (risk
management and community development) be integrated derives from the
fact that the community characteristics (e.g., community participation) and
competencies (e.g., collective efficacy) identified as influencing people's preparedness decisions develop from their everyday activities (e.g., to engage
with others for social, sporting, or political reasons) and experiences (e.g.,
collective efficacy is increased by working with others on projects that involve identifying and implementing solutions to shared local problems and
needs). Integrating risk management and community development strategies
can provide more cost-effective avenues for developing the competencies
that underpin adaptive capacity.

Risk management strategies that dovetail with community development
activities are more likely to be perceived, by community members and civic
authorities alike, as offering solutions that have immediate benefits, by facilitating the development of social capital that will show a return on investment in everyday life, and not just in the event of the occurrence of a disaster at some indeterminate time in the future. Building risk management
capacities through mainstream activities increases the likelihood that some
level of preparedness and adaptive capacity will be sustained over time. That
is, adaptive capacity can be forged and sustained through community engagement in activities concerned with identifying and dealing with local issues even if they have little or nothing to do with hazard readiness *per se.*

The fact that adaptive competencies can be developed, honed, and sustained through engaging with others to deal with more regularly occurring
community issues provides opportunities to ensure that community capacities are sustained and available to support risk management initiatives over
time. Integrating a risk management component into other community de-

velopment activities (particularly those that include collective action and problem solving elements) can develop social conduits for facilitating discussion of hazard issues, the development of problem solving competencies, and the building of collaborative relationships between communities and risk management agencies in ways consistent with the community engagement principles that underpin effective hazard preparedness strategies (see above and Chapter 6).

Given the importance of accommodating the diverse needs and interests that arise when working with multiple stakeholders, this approach could start by inviting representatives of community groups (e.g., community boards, Rotary, religious and ethnic groups, school (e.g., PTA) groups, social clubs, unions, Chambers of Commerce, industry groups, etc.) to review hazard scenarios. These discussions could identify the respective needs and expectations that would need to be included in a comprehensive risk management strategy and identify the resource and information requirements of different groups.

Discussions that focus on hazard events in this way can provide the social scaffolding necessary for community members to build the risk mitigation and protective strategies appropriate for helping them manage the personal and local implications of hazard activity. This process can help people identify the information and resources they require and structure the development of community-led risk management strategies that are consistent with the diverse beliefs, values, needs, expectation, goals, and systems within a community. A cost-effective approach would be to work with volunteers from the various groups identified. Following training and meeting their information and resource needs, these volunteers could facilitate learning in their communities using peer tutoring and collaborative learning practices. This would also provide a mechanism for providing the feedback to risk management agencies required to facilitate the progressive development of community risk management capability (e.g., Frandsen et al., 2012).

The effectiveness of these community-based activities could be increased by working with community leaders and training them to provide information and advice pertinent to the needs of their communities (Frandsen et al., 2012; McGee & Russell, 2003; Lasker, 2004). Identifying, and if necessary training, local leaders in readiness planning can provide benefits that can extend into the response and recovery contexts in which communities can find themselves should a hazard event occur. Effective, emergent leadership has been identified as crucial to developing effective recovery in events such as the Christchurch earthquake (Mamula-Seadon et al., 2012; Paton, 2012).

A variation on this approach, and one that may be useful if community leadership is lacking or needs time to develop, would involve risk and emergency management agency representatives acting as consultants to communities (e.g., acting as facilitators, resource providers, mentors, advocates, change agents, coordinators as required) rather than directing the change process in a top-down manner (Paton & Bishop, 1996). Through this process, they could assimilate and coordinate the needs and perspectives derived from community consultation, and, as far as possible, seek to provide the information and resources necessary to empower community groups and sustain self-help and resilience. By mobilizing resources intrinsic to a community, sustained preparedness is more likely to ensue. Other approaches to promoting empowerment can be found in Fetterman and Wandersman (2004). The process of developing engagement strategies can profit from learning from existing work in this area.

LONG-TERM BENEFITS OF COMMUNITY ENGAGEMENT

Risk management policies and practices built on community engagement principles can increase opportunities to appreciate public perceptions of the likelihood and consequences of hazard effects and allow for a more realistic anticipation of how community members are likely to respond to recommendations and take corrective action if misconceptions are identified. By using community engagement principles as a foundation for the development of interactive dialogue (see Chapter 6), risk management agencies can increase the effectiveness of risk management decisions and empower community members by involving them in all aspects of risk management. Engaging with all stakeholders will improve communication by developing shared understanding of the issues and implications in each community (i.e., accommodate diversity in beliefs, competencies, and needs) and so help reduce dysfunctional conflict between the community and the agencies that develop and deliver risk management activities (see Chapter 2). However, it is important to appreciate that a move to adopting community engagement strategies can pose several challenges for risk management practice and the agencies responsible for it.

For example, one consequence of increasing dialogue between citizens and agencies is a need for more open discussion about hazards and the uncertainties that surround them. Dialogue also requires more transparency in discussion about what is being done to manage risk and why (including how economic and political considerations influence decisions about public safety–Johnston et al. (2005)). Agency representatives may be unfamiliar with how to discuss risk and uncertainty with community members. This can reduce their willingness to communicate about possible risks to the public until they are sure of what the risk is (e.g., what may happen and when, etc.) and until they have identified how the risk can be effectively managed by them and by the community. However, if such a state of affairs is used to sustain top-down approaches, rather than recognizing uncertainty as a catalyst for developing two-way dialogue to develop shared understanding, this can serve to widen the gap between risk agency beliefs and goals and those of citizens (see Chapter 2). It is important to recognize that that while uncertainty is something agencies and citizens share, they do so in different ways. In order to reconcile these diverse perspectives, two-way dialogue will play a pivotal role in sharing understanding and building complementary roles in risk management. This is not the only issue likely to be novel for agencies.

The implementation of community engagement strategies may require additional training for those in risk management roles (e.g., community assessment, mentoring and advocacy skills, conflict management, etc.) and some organizational development and culture change to enhance agency capability to provide empowering community settings (Frandsen et al., 2012). This may take time. Consequently, planning processes should include assessing residual risk and strategic risk management planning to accommodate the time required to develop community and agency competencies and to plan how these development activities will be implemented over time.

The importance of being able to engage with the community derives from its role in allowing community members to actively participate in problem identification and decision-making activities in ways that can enhance members feeling as if they have some control over and involvement in the risk assessment and management processes. When the community is engaged in making risk management decisions, it is more likely to accept those decisions, accept (some) responsibility for their role in risk management, and develop a greater commitment to acting on their decisions (Paton & Bishop, 1996).

This chapter has identified and demonstrated how certain social characteristics and competencies contribute to understanding differences in levels of community hazard preparedness. There is, however, a characteristic of community life not so far covered that is less evident when analysis focuses on a single country, but which is thrown into stark relief if the level of analysis switches to an international level. This introduces another relationship characteristic with potentially significant implications for understanding how people evaluate information and relate to one another; cultural and national differences between countries. Of course, this is not just an issue when examining preparedness on an international stage: cultural diversity is an issue with significant implications for any country in which risk management processes are applied to multicultural populations (e.g., the USA, Australia).

Developing comprehensive and universally applicable theories of hazard preparedness, and developing robust risk communication and community outreach programs based on the evidence base provided by applying these theories, is a process that must consider how cultural characteristics and practices influence social interpretation and the relationships in which inter-pretation takes place. That is, they must consider how cultural characteristics and practices influence how people develop their hazard and risk beliefs and how these processes influence the deliberations required to enact beliefs in ways that assist people to anticipate, cope with, adapt to, and recover from hazard impacts. This issue is discussed in the next chapter.

Chapter 8

CROSS-CULTURAL PERSPECTIVES ON HAZARD PREPAREDNESS

INTRODUCTION

The research into hazard preparedness discussed so far in this book has emerged predominantly from Western countries, and primarily from the USA. While the experience of significant hazards in the Western countries from which much of this work has emanated means that this work does play an important role in facilitating hazard preparedness in these countries. Notwithstanding, a significant discrepancy can be discerned regarding the geographical distribution of research into risk management and preparedness and the worldwide occurrence of natural disasters and, particularly, losses from natural disasters. For example, between 1999 and 2009, some 50% of disasters worldwide occurred in Asia, with these events accounting for 90% of the deaths and 85% of the total number of people affected by disasters worldwide. Developing the capacity of people and communities in non-Western regions to mitigate and prepare for hazard events is thus of paramount importance. A logical starting point would be to import preparedness research findings from Occident to Orient (and elsewhere). However, and despite the strong empirical support for Western-derived theories, importing preparedness theory across national boundaries is not a straightforward task.

The reason for exercising some caution with regard to taking Western preparedness research onto an international stage derives from the fact that countries differ with respect to their cultural characteristics, and, importantly, differ substantially from the Western countries in which the theories described earlier were developed (see Chapters 4 to 6). For example, Asian countries, where a disproportionate proportion of disaster losses occur, tend to have a collectivistic rather than the individualistic cultural orientation that

167

prevails in the USA. If it can be demonstrated that these cultural differences have implication for preparedness and/or preparedness predictors, then a more critical appraisal of the multi-national application of preparedness theories developed and tested in Western countries is called for. Only if theories can be demonstrated to retain their predictive utility when they are tested in different countries will it become feasible to consider their more widespread application.

Before theories can be exported and imported, it is necessary to inquire whether national differences in cultural characteristics do have implications for preparedness or the preparedness process (e.g., the predictors or the relationships between predictors). If this is found to be the case it would become necessary to first assess the degree to which the theories discussed earlier (see Chapters 5–7) retain their applicability and explanatory validity in countries whose cultural characteristics differ substantially from those in which theories were developed and tested. Exploring this first requires some way of categorizing national differences.

A Cross-Cultural Perspective

Even a cursory overview of national differences reveals considerable diversity in areas such as language, dress, and cultural practices. On the face of it, this would appear to make the prospect of a country-by-country comparison an untenable task. Fortunately, studies of cultural differences have identified how these differences can be collapsed into a small set of cultural dimensions. Countries differ with regard to their relative positions on these cultural dimensions (Matsumoto & Juang, 2008). For example, this chapter alluded earlier to the cultural dimension of individualism-collectivism (I-C). This I-C dimension is one of five developed by Hofstede (2001) to describe cultural differences between countries. Hofstede's model is only one of several developed to assess cultural similarities and differences. It was selected for the work discussed in this chapter because it identified characteristics that could be implicated in the operation of preparedness processes and in determining the nature of preparedness outcomes (see below).

Hofstede's (2001) classification system affords a relatively parsimonious approach to assessing cultural similarities and differences between countries. With regard to the subject of the present chapter, it means that, rather than having to study every country, the cross-cultural applicability of theories can be assessed by comparing countries that differ with regard to their relative positions on these cultural dimensions. For example, if the explanatory utility of a theory can be demonstrated with countries that exist at opposite ends of a dimension (e.g., individualism-collectivism), it becomes possible to infer its applicability to those that fall in-between these extremes. Following research in some 72 countries, Hofstede identified how culture could be described in terms of the relative position of a country on the five dimensions of individualism-collectivism, power distance, uncertainty avoidance, masculinity-femininity, and long term orientation.

Individualism-Collectivism (I-C) describes the degree to which culture promotes, facilitates, and sustains the needs/goals of autonomous individuals over those of the group (collective). Power distance (PD) assesses the degree to which people expect and accept that power is distributed unequally and accept the authority of those defined as superiors. Uncertainty avoidance (UA) defines the extent to which members of a culture feel threatened by uncertain, unknown, or ambiguous situations. With regard to Masculinity-Femininity, a society is deemed masculine if social sex roles are clearly separated. Finally, long-term orientation is characterized by perseverance and sensitivity to status and is future-orientated.

The contents of this cultural classification identify dimensions whose nature could have significant implications for either preparedness predictors themselves and/or the relative weightings of activities within the preparedness process. For example, the relative position of a country on the I-C dimension could influence the comparative importance of person- versus group-level factors on preparedness decisions. Thus, if starting an exploration of preparedness from scratch, it would be possible to speculate that in more individualistic Western settings, individual level variables (e.g., self-efficacy, outcome expectancy) would have a greater influence on decision making than collective processes. In highly collectivistic countries, the opposite would be predicted. Similar logic can be applied to other dimensions.

It could be hypothesized that as scores on the PD dimension increase, the more likely citizens in high PD countries would be to conform with edicts from formal sources of authority regarding preparedness without question. Similarly, as the relative position of a country on the UA dimension increases, the greater would be the motivation of its populace to (unquestioningly) comply with advice in order to reduce the uncertainty implicit in their hazard-scape. Finally, the emphasis on long term orientation places on characteristics such as perseverance may influence such things as the ease with which preparedness can be sustained and become embodied in community life. These hypotheses, however, need to be put to the test to determine the degree to which they influence preparedness or, to put it another way, to test the degree of cross-cultural equivalence in theories of hazard preparedness. This issue is particularly pertinent with regard to the I-C dimension as existing theories (see Chapter 5) tend to favor intra-personal (e.g., self-efficacy, coping style) characteristics over the interpersonal processes that could be of greater importance in more collectivistic cultures. Pursuing the issue of cross-cultural equivalence has both theoretical and practical implications.

Theoretical and Practical Issues

Identifying cross-cultural equivalence in theories of hazard preparedness is of interest from the perspective of determining the degree to which the processes that underpin how people confront risk and uncertainty and respond to hazard threats represent universal social and psychological processes whose nature and action transcend cultural differences (Brislin, 2000; Norenzayan & Heine, 2005).

At a practical level, being able to demonstrate the existence of a trans-cultural conceptual framework that predicts preparedness would provide a common basis for collaborative learning and research between researchers and between practitioners across national borders. This would facilitate the ability of risk management agencies, irrespective of country, to be able to draw upon expertise gained from work undertaken anywhere in the world and to use this work to inform the development of their risk management strategies. Demonstrating equivalence would make such information available to provide a cost-effective resource for countries and communities that lack the resources to undertake this work themselves. In order to get to this position, several issues require attention.

This chapter focuses on the I-C dimension. This is the most commonly studied, and arguably the most important, dimension (Triandis, 1995). It is argued in this chapter that it is particularly appropriate for the initial exploration of the cross-cultural equivalence of preparedness theory. This argument is based on the inference that Power Distance and Uncertainty Avoidance are high in the countries being compared here (see below) and, importantly, would have more implicit influence on the adoption of preparedness measures (i.e., on the dependent variable) rather than on the predictors that are the substance of preparedness theories. For example, both can influence, either through automatic compliance (the action of high PD) or adoption to avoid uncertainty (the influence of high UA) the direct implementation of preparedness measures. However, since the dimensions, and thus the relative positions of a country on these dimensions, co-exist, further work is required to investigate these possibilities. The focus on individualism-collectivism should thus be seen only as one starting point in this process.

Thus, it can be argued that high PD and UA are cultural characteristics that could act directly on levels of preparedness in ways that could circumvent the predictors that represent the essence of preparedness theories. In contrast, I-C is a cultural characteristic that can be inferred to have direct implications for the nature of the decision process that preparedness theories (predictors) attempt to capture (e.g., the relative importance of intra- versus inter-individual and relationship variables). Consequently, the present discussion focuses on the I-C dimension.

To investigate this issue, it is first necessary to identify a theory that can be used for this comparison and so identify indicators that could be applicable across cultural divides. Because it includes both intra-personal and inter-personal variables (and so can assess more individualistic and more collectivistic processes), this chapter uses the community engagement theory discussed in Chapter 7 to provide the theoretical basis for examining the cross-cultural equivalence of preparedness. In the next section, this discussion is introduced through considering differences in how individualistic and collectivistic cultural characteristics influence behavior.

INDIVIDUAL AND COLLECTIVE INFLUENCES ON BEHAVIOR

As introduced above, the I-C dimension (Hofstede, 2001) describes how cultures can be differentiated with regard to the social and psychological bases of people's beliefs and actions (Brislin, 2000; Diener & Suh, 2000; Norenzayan & Heine, 2005; Poortinga, 1997). For example, in more individualistic cultures (e.g., Australia, USA), people act consistently across situations in accordance with a self-concept that is relatively independent of social situation and in which achieving personal goals is a prominent objective. This makes it easy to see how the kind of intra-personal processes discussed in Chapters 4 and 5 would influence preparedness. It also means that if collective action occurs, it reflects personal choice regarding levels of collaboration and cooperation rather than a cultural predisposition. In contrast, in more collectivistic cultures (e.g., Indonesia, China), actions in most domains of daily life are underpinned by culturally-embedded beliefs that are reflected in shared purpose and activities that align with social norms to achieve collective goals through engaging in activities related to future goals that emphasize social relations (Diener & Suh, 2000; Jang & LaMendola, 2006; Triandis, 1995).

In light of these cultural differences, it becomes pertinent to ask whether there are grounds for believing that collective processes influence hazard preparedness in relatively more individualistic cultures and whether individual-level interpretive processes and beliefs influence preparedness in collectivistic countries. Only if affirmative answers to these questions are forthcoming will it become feasible to explore equivalence and then assess whether a valid cross-cultural theory can be developed. This chapter discusses first reasons why affirmative answers to these questions might be expected.

Foundations of Cross-Cultural Equivalence

The potential for social factors to influence beliefs and actions in Western, more individualistic countries, was introduced in Chapter 6 (symbolic interactionism) and Chapter 7 discussed empirical evidence of how, when faced with uncertainty, people turn to others who share their interests and values to help them reduce uncertainty and decide how to manage their risk (Earle, 2004; Lion et al., 2002; Paton, 2008). Family and members of the communities (e.g., workplaces, social and sporting clubs, churches, etc.) with whom people interact regularly are prominent sources of this collective assistance. Thus collaborative processes can be implicated as playing significant roles in the social construction of risk beliefs and preparedness decisions in members of individualistic cultures. This makes it feasible to consider comparing individualistic cultures with their collectivistic counterparts (for whom social context is, ostensibly, a more implicit driver of decision making and actions).

The next question is whether a basis for comparison exists in the other direction. In individualistic cultures (e.g., USA, Australia), it is not surprising that individual-level variables (see Chapters 4 and 5) have been found to be prominent predictors of hazard preparedness. However, can the same be said for their collectivistic counterparts? The answer would appear to be a (qualified) yes.

Individualistic traits are being recognized for their potential to influence risk management choices in members of collectivistic cultures such as Japan (e.g., Bajek, Matsuda, & Okada, 2008; Childs, 2008; Nakano, 2005; Paton, Bajek, Okada, & McIvor, 2010; Tanaka, 2005; Tatsuki, 2000). The existence of empirical evidence indicating that person-level variables (e.g., outcome expectancy) can influence the choices made by citizens in collectivistic cultures offers additional justification for undertaking a comparison of the community engagement theory. An argument for expecting equivalence between members of individualistic and collectivistic cultures can also be made on the grounds of anticipated similarities in how people acquire the information required to help them deal with uncertainty.

The infrequent and complex nature of natural hazard events in all countries and the limited (if any) opportunities their citizens have to gain experience of either the hazard consequences they could experience or the effectiveness of mitigation measures they are being asked to adopt means that the climate of risk communication in all cultures is characterized by considerable uncertainty. Under this circumstance, if people are to identify the kinds of hazard consequences and demands they may have to contend with and find out what they might do to manage their risk, members of all cultures have to rely, at least to some extent, on information from expert scientific and risk management sources (e.g., through public education programs, attendance at public meetings, etc.). This makes it pertinent to include the kind of community-civic agency relationships discussed in Chapter 7 (e.g., the role of empowerment and trust) in a cross-cultural comparison of hazard preparedness.

The above discussion suggests that irrespective of the cultural characteristics of the population being investigated, there exist grounds for believing that people's decisions about hazard preparedness result from interaction between individual beliefs, collective processes and competencies, and the quality of the relationship between people and the civic agencies. This provides additional justification for using the community engagement theory (Chapter 7) for a cross-cultural comparison.

To test for equivalence, testing was done by comparing countries that differ substantially on their relative positions on the individualism-collectivism dimension. The next section discusses the findings from comparing volcanic preparedness processes in New Zealand and Indonesia and from a comparison of earthquake preparedness in New Zealand and Japan (Paton, 2008; Paton, in press; Paton et al., 2010). See Chapter 7 for details of the theory.

Cultural Equivalence in Predictors of Hazard Preparedness

The relative I-C scores of each of the countries included in this comparison are described in Table 8.1. Conducting a cross-cultural comparison introduces another challenge for the researcher. That is, the need to translate questionnaire items into other languages in ways that ensure that the underlying meaning of the items in the scales being used to assess variables is comparable. The research on which the discussion in the present section is based used the approach recommended by Brislin (1986). First, the English version was translated into each language. Second, that version was translated back into English by another independent translator. The original versions and back-translated versions were compared, examined for meaning errors, and corrections were made as required to ensure equivalent content. Tests for the face validity of the scales were also conducted and supported the use of the variables included in the theory.

It was also necessary to ensure some comparability with regard to the dependent variable. To achieve this, intention to prepare was used as the dependent variable. Intentions can accommodate hazards with diverse pre-

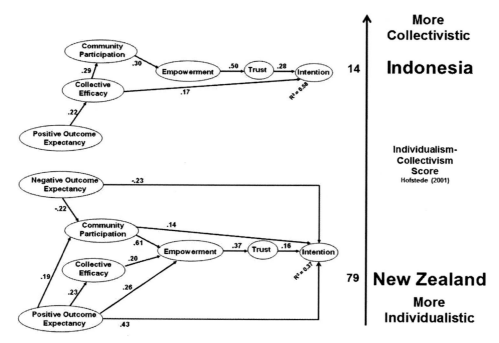

Figure 8.1. A comparison of the community engagement theory for volcanic hazards.

paredness needs (e.g., the measures required to adapt to volcanic hazards differ in several ways from those relating to earthquake hazards). Intentions can also facilitate cross-cultural comparison of communities that differ with regard to characteristics such as hazard history, culture, public education strategies, and recommended preparedness measures. The comparisons are summarized in Figure 8.1 (volcanic) and Figure 8.2 (earthquake). What implications do these findings have for understanding cross-cultural equivalence in predictors and processes of hazard preparedness?

Theory Equivalence

Cross-cultural equivalence was evident for both hazards (Figures 8.1 and 8.2). There were some path differences, but there was good comparability regarding the variables supported by the analyses and in the relationships between variables. For volcanic hazards, personal beliefs (outcome expectancy variables) interacted with community processes (community participation and collective efficacy) and institutional (empowerment and trust) factors to influence intentions to prepare. A similar picture was evident for earthquake hazards.

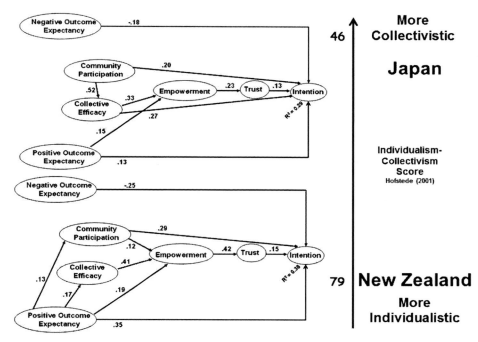

Figure 8.2. A comparison of the community engagement theory for earathquake hazards.

This comparison (Figures 8.1 and 8.2) supports the view that, irrespective of culture (at least with regard to their relative position on the individualism-collectivism dimension), beliefs regarding the efficacy or otherwise of preparedness measures influences how people make hazard preparedness decisions. This comparison also supports the position that the more citizens are able to collectively formulate their risk management needs and strategies (assessed by collective efficacy) and the more they perceive their needs as having been met through their relationship with civic agencies (assessed using empowerment), the more likely they are to trust agencies and the information they provide, and to use information from these agencies in their hazard preparation adoption decision making.

Demonstrating some degree of cross-cultural universality in how people deal with the risk and uncertainty associated with, and make decisions about, hazard preparedness has implications for risk management and for learning about how processes such as community participation develop and act with regard to hazard preparedness. Finding this level of equivalence is consistent with Triandis et al.'s (1986) work that introduced the concepts of allocentrism and idiocentrism to accommodate the fact that both individual and social factors influence behavior in all cultures through the enactment of cultural beliefs in everyday life, though they may differ in the relative importance of individual versus collective influences at a national level. The existence of some measure of comparability across cultures in the process of prepared-ness offers support for the idea of using a theory to increase the scope for international collaboration in both preparedness research and intervention design.

The demonstration of a degree of equivalence in the process used by people to make preparedness decisions does not mean that people in differ-ent countries participate in the same way or develop competencies such as collective efficacy in the same way. Additional insights into the nature of the social influences on preparedness can be gleaned from exploring how the content of the process (e.g., how participation occurs, how people collabo-rate to deal with challenging circumstances) takes place in different countries. This issue is discussed in the next section.

CULTURE-SPECIFIC INFLUENCES
ON THE PREPARING PROCESS

An exploration of the nature of social relationships in different countries reveals how cultural specific processes could affect how the theory is enact-ed, particularly with regard to the factors such as participation, collective effi-cacy and empowerment. For example, community participation in Japan may have been influenced by Jishubo (autonomous community-based orga-nizations for disaster prevention) and its role in facilitating community mem-ber's satisfaction with and the quality of their learning about hazard man-agement (Bajek et al., 2008). Furthermore, the effectiveness of Jishubo as a risk management mechanism is influenced by its operating under the aus-pices of an approach to societal governance that is unique to Japan.

Local government plays a direct role in organizing community partici-pation in disaster prevention in Japan, with much disaster prevention-related policy there being implemented through the governance scheme of Chonaikai (Bajek et al., 2008). In this regard, evidence for the effective role of mechanisms such as Jishubo, particularly if people are active volunteers (Bajek et al., 2008), provides tangible evidence of the benefits accruing from the presence of empowering links between communities and civic agencies within risk management strategies. Thus, through Chonaikai in general and Jishubo in particular, it is possible to identify culture specific content for the community participation (e.g., Jishubo affords opportunities for community members to participate in social activities that align with disaster prevention goals) and collective efficacy (e.g., Jishubo affords opportunities for community members to work together to define and resolve local risk management issues) measures in the Japanese sample. Furthermore, there is evidence of a relationship between involvement in Chonaikai-organized events, community trust in civic agencies and preparedness (Bhandari, Okada, Yokomatsu & Ikeo, 2010), further reinforcing understanding how culturally specific processes could exercise some influence on the action of core theory variables such as trust (see Chapter 7).

Similar culture specific processes can be identified in Indonesia; the traditional practice of gotong royong (mutual help). Gotong royong describes a culture-specific process that underpins a prevailing sense of mutual responsibility for acting to promote the well-being of a community in areas of Indonesia. It thus describes a cultural mechanism that promotes participation in ways that involve people working collaboratively to deal with shared issues and opportunities. From this description, it is possible to anticipate how this could represent the source of the community participation and collective efficacy variables in the theory.

These examples illustrate how knowledge of culture-specific mechanisms can offer insights into how social processes and competencies such as community participation and collective efficacy contribute to how preparedness beliefs develop and are enacted. The analysis of how variables such as community participation and collective efficacy are enacted in different countries draws attention to a need to clearly differentiate the process of preparedness and the specific content of the process. The preceding discussion of culture-specific processes demonstrated how mainstream cultural practices could have positive influences on hazard preparedness. This need not always be the case.

One example of how implicit cultural characteristics can reduce the likelihood of people preparing comes from Hawai'i. In the context of the possibility of using engineering solutions to divert lava flows from future volcanic eruptions in Hawaii, Gregg, Houghton, Paton, Swanson, and Johnston (2008) examined the role of cultural sensitivity associated with Hawaiian beliefs in Pele, the Goddess of volcanoes. Because native Hawaiians believe that interfering with lava flows disrespects the wishes of Pele, this belief influenced their attitudes to using an engineering solution to mitigate lava flow hazards.

Gregg et al. (2008) compared levels of support for engineering mitigation measures (building walls to divert lava flows, bombing lava flows) amongst those identifying with Hawaiian ethnicity with those identifying with other ethnicities. They found that people whose cultural beliefs derived from identifying with Hawaiian ethnicity (e.g., belief in Pele) were significantly more likely to object to using or even supporting the use of those mitigation actions that were inconsistent with their cultural beliefs.

These examples illustrate the importance of accommodating cultural beliefs (e.g., reflecting spiritual or environmental beliefs) in risk management planning and in research. Failure to accommodate cultural beliefs, needs, and expectations or promoting and implementing actions that are inconsistent with cultural beliefs, could have the effect of reducing trust in the sources of information, advice, and resources people rely on when dealing with complex and uncertain events and adding unnecessary complications to the risk management process. In contrast, accommodating them can expedite the risk communication process.

CULTURAL INSIGHTS INTO COMMUNITY DEVELOPMENT

The contents of the first part of this chapter support taking the position that, irrespective of the position of a country on the I-C dimension, the characteristics (e.g., community participation) and competencies (e.g., collective efficacy) that develop through accumulated experience and interaction in people's social context influence both how people interpret environmental uncertainty and make decisions about preparing for nature hazard events. The research discussed here reinforces the importance of accommodating social context influences in theoretical formulations of how risk beliefs develop and are enacted as preparedness strategies (see also Chapter 6). The work discussed above also demonstrates that people's perception of the quality of their relationship with scientific and risk management agencies (e.g., whether it empowered them, whether they trust them) appears to be a culturally universal aspect of hazard preparedness decision making, at least with regard to infrequently occurring hazards.

This chapter also reiterates how (pre-existing) mainstream (i.e., derived from everyday experiences) community competencies and characteristics can influence hazard preparedness (see Chapter 7). This provides additional support for the potential benefits that could accrue from developing risk management strategies by integrating them with mainstream community development activities (Anckermann et al., 2005; Paton, 2008; Paton & Jang, 2011; Rich et al., 1995). Risk management strategies that dovetail with community development activities can capitalize on community engagement activities concerned with identifying and dealing with local issues (even if they have little or nothing to do with hazard readiness per se) to develop the processes and competencies that can subsequently be used to support preparedness.

The culture-specific issues discussed above should not be ignored. A failure to accommodate cultural issues could increase vulnerability (e.g., because people will not support or adopt measures that are culturally inappropriate) or increase the risk of community fragmentation (e.g., fuelling disagreement amongst members of the same community, reducing future levels of community participation on hazard issues). Proposing measures that conflict with cultural beliefs or practices could increase distrust of civic authorities responsible for risk management, result in risk management policies and strategies being inconsistent with social justice principles, and increase the likelihood of strategies and ideas being ignored. In other words, a failure to accommodate cultural diversity could have the counterproductive effect of reducing community resilience and increasing vulnerability.

Consistent with the issues raised in Chapters 6 and 7, this chapter reiterates how the quality of the relationships between people and societal-level institutions and agencies does influence hazard preparedness. While discussion so far focused predominantly on societal or civic source of risk information, this is not the only societal-level influence on people's recovery. Another influence on people's adaptive capacity and resilience relates to the role businesses play in maintaining the economic and social fabric of the community and in assisting people to recover from hazard impacts. This makes business preparedness an important influence on people's adaptive capacity. Its implications in this regard are the subject tackled in the next chapter.

Table 8.1.

Location	Individualism Score
New Zealand	79
Japan	46
Indonesia	14

Chapter 9

BUSINESS PREPAREDNESS

INTRODUCTION

Hazard preparedness has been defined in this book as a process intended to facilitate the ability of people and communities to cope with, adapt to, and recover from the consequences of large-scale hazard events. The discussion of how preparedness needs change as people transition through the phases of disaster (impact, response, recovery—see Chapter 3) introduced how, over time, the effectiveness of people's recovery is increasingly influenced by how they engage with or relate to civic agencies. This chapter adds a need to consider how understanding the relationship between businesses and their employees and between businesses and communities can contribute to developing a comprehensive understanding of preparedness. A significant reason for including community-business relationships stems from the fact that community recovery and business recovery are, at least in part, inter-dependent.

Most research on disaster impacts focuses on what people will have to contend with as a result of direct hazard impacts. The inclusion of business preparedness in a book primarily concerned with community preparedness draws attention to the fact that the economic and business consequences of large-scale hazard impacts are issues people should prepare for.

Well prepared community members recover more effectively and, in doing so, increase the likelihood of their being available to support business and economic recovery (as employees and customers). Businesses that develop the capacity to recover can continue to provide employment and safeguard the livelihoods of community members (both directly through increasing the likelihood that people can retain their jobs and indirectly through providing employment via response and recovery-related activities) and sustain the social and economic vitality of areas affected by disaster. If businesses disappear, people have to travel further for goods and services and the long-term sustainability of community life can be threatened or compromised. Business preparedness, particularly for small businesses, is thus vital to sustaining community life. Or, to put it another way, the effectiveness of people's preparedness will be muted if businesses have not prepared in ways that allow people to regain normal functioning, and the effectiveness of business and economic recovery will be subdued if people have not taken steps to increase the likelihood of their being available to participate in community and economic life. Business recovery principles are also relevant for organizations in the non-commercial and government sectors (e.g., minimizes disruption to societal services (e.g., social and welfare) and facilitate effective service delivery under exceptional circumstances). The process of pursuing the business or organizational contribution to comprehensive recovery falls under the heading of business continuity planning and management.

This view thus portrays people, communities, and businesses (and other government organizations such as welfare and social services agencies) as playing complementary roles in risk management, particularly with regard to their roles in facilitating effective, comprehensive recovery following disaster. The preceding chapters have focused on what people and communities should do to prepare for hazard events. This chapter considers what businesses and public institutions (e.g., public services, welfare agencies, etc.) have to do to prepare to ensure their ability to contribute to comprehensive recovery.

Just like its community counterpart, business preparedness is important for all businesses operating in areas susceptible to natural hazards. Natural disasters can have a huge impact on businesses. For example, the Loma Prieta (San Francisco) earthquake in 1989 resulted in 60% of local businesses being destroyed or closed temporarily. Businesses that close for even a short time are placed at a major competitive disadvantage relative to those that can stay open. For example, the businesses that suffered the most disruption in the Northridge, California earthquake were least likely to recover in the subsequent three years (Dahlhamer & Tierney, 1996).

However, and just like its community counterpart, business disaster preparedness is not a widespread practice. A survey in 2008 by FM Global identified that some 96% of senior executives stated that their business was exposed to natural hazards such as hurricanes, floods, and earthquakes. However, fewer than 20% indicated that their businesses were concerned about the potential of such hazards to adversely affect their bottom line. That this is the case is not really unexpected.

Previous chapters discussed how people can accept their risk but still fail to do anything about it. Business owners and executives are people, too, and their beliefs and actions are influenced by the same processes identified as affecting people in everyday life (e.g., see Chapter 4 and 5). However, the lack of business preparedness, and motivation to prepare, takes on additional significance as a result of its (avoidable) implications for adversely affecting economic vitality and people's livelihoods should a disaster occur.

The importance of taking steps to manage business risk from disaster was illustrated by Levene (2004). Levene discussed how a lack of business preparedness accounted for about 25% of the $40 billion lost as a result of the September 11 terrorist attacks in New York. Levene called for greater emphasis to be paid to developing the capacity of businesses to adapt to interruption to business activity from disasters. In support of this call, he cited evidence that an estimated 90% of medium to large companies that can't resume near-normal operations within five days of a crisis face substantially increased risk of going out of business within five years of experiencing a disaster. These problems may be considerably more acute for small businesses. So, what might businesses and their employees have to contend with?

Business Impacts

Just as the earlier discussion of community preparedness started by considering the demands people could have to contend with, this is a good place to start a discussion of business preparedness. A disaster can create significant production and distribution problems and affect patterns of the consumption of goods and services. It will also create several issues for employees.

Production problems can arise for several reasons, including loss and/or inability to access premises; loss or damage to equipment, goods, and materials; inability to acquire goods and materials; employee absences; loss of power; and loss of critical data. The distribution of goods and services can be disrupted by transportation and telecommunications problems, damage to vehicles, and failure of interdependencies between suppliers and businesses. For instance, in the aftermath of the earthquake in Kobe, car companies like Toyota experienced problems with "just-in-time" production processes as a result of knock-on effects from sub-contractors being unable to supply essential goods.

Similar issues emerged after the 2011 earthquake and tsunami in Japan, following which interruptions to business activity were expected to persist for several months. Where disruption to transportation occurs, it can be more expensive. Drawing on the Kobe earthquake experience, transportation costs in Japan increased by some 50%, and this resulted in an increase of 10% in prices of goods. Patterns of consumption can change and differentially affect some businesses more than others during the recovery period. People are more likely to be concerned with meeting basic needs, resulting in reductions in consumption of non-essential goods and services (e.g., restaurants, cinemas). However, in some cases at least, business impacts can include more positive elements.

Businesses meeting basic and recovery/rebuilding needs can, if in a position to continue functioning, take advantage of increased demand for, for example, building and infrastructure reconstruction and accommodation. The latter, particularly for hotels, motels, and guesthouses, also raises issues about business preparedness and tourism. While not tackled here, this is an important area where future research is needed.

A well-prepared business may also be able to experience growth following disaster. They may be one of the functioning businesses in the area and benefit from increased demand for goods and services. Businesses whose employees possess skills that will be in high demand in the response and recovery environment (e.g., those in trades such as building and plumbing that are essential to rebuilding) can experience a period of higher demand and higher wages.

Given the importance of business activity for economic and social well-being, it is also important to acknowledge the role that business continuity management plays in facilitating community and societal recovery from disaster. A key area of business contribution to societal recovery coincides with the need to accommodate the impact of disaster on employees.

Employee Impacts

Employees may be unable to work if the business cannot function or cannot recover. For example, the Kobe earthquake resulted in some 4,500 redundancies. Employees may be prevented from working if they are temporarily displaced to safer areas or if damage to the business premises or its immediate environment prevents them working there. In Christchurch, the CBD was still closed as a red zone for demolition in late 2012—more than 18 months after the 2011 earthquake. Disruption to business from hazard activity may mean that even if employees retain their jobs, they may not be paid for some time.

Businesses have to contend with possible temporary loss of employees (e.g., employees who are unable to get to work, injured, or psychologically incapacitated by stress and traumatic stress) or sometimes permanent loss of employees (e.g., death of employees, permanent re-settlement of workers to other areas). The availability of employees is also a function of levels of employee home and family preparedness. Well-prepared employees will be able to adapt more readily to any disruption caused by the disaster and better able to return to work. This means that one component of business continuity planning will be to encourage and guide employee and household planning in order to increase the likely availability of employees in the event of a disaster. The opposite is also true.

Employees, as part of their general preparedness, should also inquire as to how prepared their employer is and what plans are in place for business recovery and interruption. For example, it is useful for employees to know things such as what might happen to salaries and wages if employment is disrupted or the implications of having to operate from alternative premises, etc. By inquiring about these as part of their preparedness, employees put them themselves in a better position to anticipate such secondary consequences as being without income or having to relocate. By anticipating them, they can take steps to prepare for them and put themselves in a better position to respond (rather than having to react) should such conditions eventuate. These examples provide further illustrations of how employee and business interests can be interdependent and, consequently, influence community recovery.

Business and Community

Businesses constitute relational communities (see Chapter 6) and the social settings they create can provide structured and informal social support and help employees access informational, tangible and emotional support, monitor mental health issues, and provide access to professional support (Paton, 1997). While this can be an extension of the employee assistance programs large businesses often provide, for smaller enterprises, it may be a need that is met through liaison with welfare and social service groups. The development of these relationships needs to be developed in advance and included in business planning. Catering for the livelihood and well-being needs of employees is good for both business and social recovery. Well-prepared businesses provide opportunities for employees to regain a sense of control in what may otherwise be an environment characterized by chaos. Businesses that are capable of continued functioning can contribute to community recovery in other ways.

Community recovery can be facilitated by businesses providing cash donations or loans or by businesses using their expertise and connections to run fundraisers, and their being responsive to community needs during the recovery period (Paton, 2012). They can also donate technical expertise, equipment, and, where the business has direct relevance for community recovery (e.g., builders, counselling), services that assist relief and recovery efforts. In order to realize these benefits, businesses need to be prepared and their employees need to be prepared.

It is thus important that businesses anticipate what could happen, plan accordingly, and develop their capacity to implement the plan. How businesses manage risks and develop their capacity for continued operations during and after disasters is crucial for their survival (Elliott, Swartz, & Herbane, 2002; Rose & Lim, 2002). Doing so can be beneficial. For example, the investment bank, Morgan Stanley, was the largest tenant in the World Trade Center in New York. They realized after the previous attack on the Center in 1993 that they were very vulnerable to future terrorist attacks. Accordingly they established contingency and continuity plans which were tested rigorously and regularly. As a result, the company began evacuating its employees to its three recovery sites one minute after the first plane flew into the World Trade Center on 9/11 and they lost only seven employees (Coutu, 2002). An appropriate starting point for this discussion is to ask what do businesses do, or not do.

BUSINESS PREPAREDNESS

As with citizens, not all businesses prepare and, as with citizens, it is important to account for the reasons why some businesses prepare and others in similar environmental circumstances do not. Dahlhamer and Souza (1995) examined several predictors of company preparedness in the USA: business size, business age, type of business, whether it was individual or a franchise, and whether it was owned or leased. They surveyed 27,000 businesses in Shelby County and 14,000 in Polk County, using both phone calls and mailouts. The response rate was 40% and the majority of participants in the survey were small businesses.

A regression analysis showed that the business characteristics model (size and type of business) was a significant predictor of preparedness, explaining about 15% of the variance in the level of preparedness. The size of the business, defined in terms of the number of full-time employees, was the strongest predictor. Larger companies were more prepared (see also Webb, Tierney, & Dahlhamer, 2000), and more likely to have a business continuity plan (see below), whereas many smaller companies did not. In one of the two counties, the type of business was a predictor, in that finance, insurance, and real estate sectors were more prepared than other sectors. This may reflect the likelihood that businesses in these industry sectors have financial risk management procedures in place, making it easier for them to adapt plans to accommodate risk from other sources. Firms that were part of a national chain of businesses were more prepared than firms run by sole proprietors.

The authors concluded from their findings that incentives and regulation are necessary to achieve high levels of preparedness as the level of voluntary preparedness in businesses was so low.

Research has also shown that businesses that have had previous experience of a natural disaster are more prepared. Webb et al. (2000) compared businesses in three areas that had experienced large disasters (the Loma Prieta earthquake; the Northridge earthquake, and Hurricane Andrew in Florida) with businesses in areas where there had been no such disaster (e.g., Des Moines, Memphis). In some regions, preparedness also related to the type of business. For example, in Northridge and Memphis, the finance, insurance, and real-estate sectors were more prepared than retail and service areas (see above). This finding highlights the importance of tailoring outreach activities to the needs of different sectors in order to target their specific needs, rather than sending the same undifferentiated message to all types of companies (see also Chapter 7).

Yoshida and Delye (2005) examined predictors of Florida businesses carrying out mitigation actions, having a business continuity plan, and purchasing insurance. A regression analysis found three significant predictors of businesses' decision to employ mitigation measures. The strongest predictor was access to expertise such as a structural engineer; the second predictor was the type of business.

Businesses in the education, social services, finance, insurance, and real estate sectors had higher levels of preparedness and more likely to have business continuity plans than other businesses. In some cases, this difference may be because many of the companies in these sectors are required by their national or state professional organizations to have a business continuity plan. In contrast, businesses in engineering, architecture, and accounting had carried out more mitigation measures. The third significant predictor was their perceived exposure to natural hazards–a measure of companies' perception of the risk. As in other studies, small companies were less prepared, as were businesses that were run from the owner's home. In addition, many more companies had commercial property insurance than had flood or business interruption insurance. This pattern reflects Slovic's (1986) finding that people's perception of the relative risk or frequency of hazards often does not parallel the objective risk of those hazards occurring. Consequently, business education should target small businesses, those run from homes and those types of business where preparation is lower. Attending to this issue takes on greater importance in light of the greater likelihood of small business going under in the event of a disaster (see above).

In response to a question as to what information would be most useful to them, the highest rated options in Yoshida and Delye's (2005) study were a do-it-yourself business disaster continuity plan workbook, a list of organizations that provide mitigation information, a CD for business vulnerability assessment and continuity planning, and a list of publications on mitigation measures. Groups such as Chambers of Commerce and industry organizations can assist small businesses to consider the kinds of non-routine contingencies they may have to contend with and help them develop plans. Other organizations, such as Rotary, may also be able to provide experienced people to mentor small business owners in disaster business continuity management. To further this process, organizations need to know of the types of preparedness available to them.

Types of Preparation: Mitigation Versus Survival

Actions taken to prepare for hazards such as earthquakes include survival actions (e.g., purchasing a medical kit, battery radio, food and water, etc.) and actions that mitigate damage from hazard activity (see Chapter 3). A large proportion of the deaths and damage in disasters results from a failure to take adequate mitigation actions (which save lives and minimize property damage), such as strengthening buildings (Russell et al., 1995). Research shows that both individuals and businesses undertake significantly fewer actions to mitigate damage than survival preparedness actions (Russell et al., 1995; Webb et al., 2000; Yoshida & Delye, 2006). This issue is particularly relevant for the resilience of small businesses. More broadly, businesses are more likely to perform measures that impact on saving life, and are less expensive and less complicated to perform, in the short term rather than measures aimed at facilitating business survival in the long term such as measures to mitigate damage. Webb et al. found that few businesses had made plans to move to another site if this proved to be necessary after a disaster.

Partly in response to observed low levels of uptake of mitigation measures, Yoshida and Deyle (2005) examined what factors influence small businesses adopting hazard mitigation measures. A distinctive aspect of this study is that it asked respondents what access they had to four different types of mitigation experts: insurance manager, structural engineer, businesses continuity specialist, and disaster recovery specialist. The most common way of engaging expert use was access through consultants rather than employing specialists. Those businesses that had access to all four types of experts were larger and were more likely to view themselves as vulnerable to adverse hazard effects.

A regression analysis found that access to expertise was a significant predictor of businesses' decision to employ measures to mitigate damage, as were perceived exposure to natural hazards and type of business. Businesses in education, social services, finance, insurance, and real estate were more likely to have a business continuity plan (BCP) in place, and perceived exposure to hazards was positively related to structural mitigation and insurance purchase. Though this chapter portrays businesses as the focus of interest, it is important to understand that preparation decisions are often taken by individuals (e.g., business owners), particularly in small businesses. This may result in their decision making being susceptible to the kinds of biases discussed in Chapter 4.

Other research suggests that the difference in the extent to which people undertake survival and mitigation actions may occur partly because mitigation actions are perceived as having a higher cost (Webb et al., 2000). Yoshida and Delye (2005) similarly found that the three most commonly adopted measures in a sample of businesses in Florida, USA cost less than $500. However, this difference in performance of the actions is maintained even with regard to the performance of mitigation actions that are low in cost, such as fitting computer restraints or securing bookshelves (McClure et al., 2007). For example, in the 1989 Loma Prieta (San Francisco) earthquake, the strongest predictor of business resilience was whether businesses had undertaken mitigation actions and in particular whether companies had used computer restraints or back-ups, which involve a relatively low cost (Yoshida & Delye, 2006). This discussion prompts further consideration of how people appraise the costs and benefits associated with preparedness decisions, and how this appraisal influences their motivation to act.

Motivation: Costs and Benefits

Although many strategies to get people to prepare for hazards involve communicating information about the risks and hazards, information on its own, as discussed in the preceding chapters, is often insufficient to get people or companies to prepare. Many people also require motivation, which may take the form of incentives or rewards or regulation such as legislated building standards.

In a recent study of business preparedness for earthquakes, the participants were given a 56-page Emergency Management booklet about company preparedness and a brochure about computer restraints (McClure et al., 2009). The participants were also asked how useful they found the booklet and brochures, and they almost universally said the information was very useful and informative. However, many of them did not use this information to increase their preparedness, and the preparedness actions that companies did perform bore no relation to their rating of the usefulness of the brochures. This finding supports the view that to get companies or individuals to spend time and money to take action, information needs to be reinforced by incentives and/or regulation.

Motivation is often construed in terms of the costs and benefits relating to a given action. A cost-benefit analysis is often employed in businesses to assess whether the expected benefits from a proposed action exceed its expected costs. If the total value of the costs exceeds the total value of the benefits, the relevant action should not be taken.

In targeting hazard preparedness in terms of costs and benefits, it is important to counter the perception that only major expenditures are useful in mitigating damage from huge events like floods or earthquakes. In the Loma Prieta earthquake, the strongest predictor of business survival was whether or not companies had a computer lock. Businesses that had no computer locks lost all their data and records, and took months to regain some or all of this data, often going bankrupt as a consequence. In contrast, companies that had secure computers were able to function the following day or week, even if in some cases they had to shift premises due to damage to their building. In this case, the very small cost of the computer security had huge benefits. With large companies, this security involves servers and other backups rather than computer locks, but the principle still holds, because these security measures are relatively inexpensive. Thus it is important to communicate that large benefits do not necessarily entail large costs.

A recent study examined the relation of cost to the citizens' performance of a range of survival and mitigation actions (McClure, Fischer, Charleson, & Spittal, 2009). In this study, the mitigation actions were divided into two groups: actions that affect damage to contents, such as securing shelves, and actions affecting damage to or collapse of the building, such as securing foundations. Households and businesses estimated the cost of a sample of three types of actions: survival actions, actions to mitigate contents damage, and actions to mitigate structural damage.

The results showed that households completed fewer of both types of mitigation actions than survival actions, whereas businesses completed fewer structural mitigation actions than contents damage mitigation actions and survival actions. Participants also completed fewer actions they rated as more expensive than less expensive actions. With households, there was a modest but significant relationship between the perceived cost of four of the 12 actions and the probability of carrying out the actions. With companies, there was no such relationship between actions and cost.

The businesses in this study set goals of preparedness actions to complete three months later. Again, cost did not have a significant relation to either the actions that they selected to complete or to the actual completion of the actions. On a measure of attributions for why they did not carry out the given actions, the perceived cost of the actions was only the fourth, fifth, or seventh most common attribution for both households and businesses not completing the actions (the exact ranking depending on the type of action and the sample: households or businesses). The three attributions that were more frequent than cost for both samples were: First, I/we haven't thought about it; second, it's not a priority; and third, the belief that it would make no difference (e.g., negative outcome expectancy). These results suggest that the cost of preparation actions is one factor hindering preparedness actions, but that these other reasons people gave for not preparing are more important. These observations highlight the need for more systematic considerations of disaster response needs. In a business context, this more systematic consideration takes the form of business continuity planning and its translation into preparedness.

BUSINESS CONTINUITY PLANNING AND PREPAREDNESS

Shaw and Harrald (2004) define business continuity planning and management as comprising those practices that focus and guide the decisions and actions required to prevent, mitigate, prepare for, respond to, resume, recover, and restore business activity should the business experience large-scale hazard events. Business continuity planning (BCP) develops the ability of business systems and employees to act in ways that can sustain activity in the face of significant disturbances and to recover as quickly as possible in the event of loss and disruption to business activity. Developing this capacity gives businesses the capability to draw upon their resources and competencies to manage the demands, challenges, and changes encountered prior to, during, and after hazard events (Comfort, 1994; Paton, 2006). Preparing a business to be able to cope with and recover from the consequences of hazard activity requires the systematic appraisal of the conditions that could necessitate change, and the development of the systems and staff competencies capable of facilitating continuity of business activity under atypical crisis conditions.

Pursuing this objective requires businesses to attend to several issues. First, it is important to implement and develop appropriate survival and mitigation actions (see above). Second, it requires that management and information systems are available (by safeguarding existing systems and/or arranging for substitutes) to facilitate continuity of core business operations (Davies & Walters, 1998; Duitch & Oppelt, 1997; Lister, 1996). Third, it requires the development, testing and regular up-dating of management systems and procedures designed specifically for managing a crisis and for managing the transition between routine and crisis operations in the context of the kind of losses and disruption that the business may have to contend with in the event of their experiencing a large-scale natural disaster (Paton, 1997; Shaw & Harrald, 2004).

Disasters associated with natural hazard activity, such as that likely to accompany large-scale earthquakes, volcanic eruptions, or floods, represent the upper end of the scale of events that need to be considered within the business continuity planning process (Reiss, 2004). While businesses often prepare for small-scale disruptions or crises (e.g., loss of power supply for a few days), planning for such low-level events will not confer upon a business a capacity to respond to more significant events. Qualitative differences in the nature, scale, and duration of the impacts and consequences associated with large-scale hazard events mean that planning for routine losses represents an inappropriate basis for disaster business recovery planning. However, by planning for large scale disasters, BCPs are able to accommodate the impact of lesser events (e.g., loss of utilities). Some of these differences are illustrated in Table 9.1.

When planning for large-scale disasters, businesses must consider how they will deal with, for example, conducting core operations away from their HQ. Even if able to operate from their normal premises, planning requires that they anticipate having to function with prolonged loss of or disruptions to utilities (e.g., power, water, gas), dealing with disruptions to patterns of supply, distribution, and consumption, and ensuring that appropriate crisis management systems and procedures are in place, and that staff will be able to work over the prolonged period (possibly over several months) of response and recovery. Planning also needs to consider issues such as dealing with casualties and deaths amongst staff and reconciling work with the family needs and concerns of staff during response and recovery. The implementation of these plans requires that consideration be given to business continuity management.

Business Continuity Management

Business Continuity Management (BCM) is a proactive, holistic, and structured management process that incorporates risk identification, assessment, and management; the development of disaster recovery plans and procedures; training, exercising, and using feedback from exercises to promote the iterative development of plans and capabilities. BCM is built around understanding what the business must achieve (its critical objectives), identifying the barriers or interruptions that may prevent their achievement if a business experiences large-scale natural hazard consequences, and planning

Table 9.1.
DIFFERENCES BETWEEN EMERGENCY AND DISASTER OPERATING
CONTEXTS. ADAPTED FROM AUF DER HEIDE (1989) AND PATON (1997).

Interaction with familiar people	Interaction with unfamiliar people (different organizations and agencies)
Familiar roles, task, and procedures	Unfamiliar roles, task, and procedures
Intra-organizational coordination needed	Intra- *and* inter-organizational coordination needed
Environment and infrastructure (roads, telephones, and facilities, etc.) intact and useable	Environment and infrastructure damaged and possible unusable (e.g., roads blocked/jammed, telephones overloaded, or non-functional, facilities damaged)
Communication adequate	Communication infrastructure lost or damaged and communication inadequate
Communication and decision making primarily intra-organizational	Need for inter-organizational communication, information sharing and decision making
Use of familiar terminology and procedures	Working with others and using unfamiliar terminology and procedures that require dedicated disaster continuity procedures and training
Management structures and response capacity adequate to coordinate resources and respond	Management and response capacity exceeded by quality and quantity of demands
Buildings intact and functional	Buildings damaged and necessary to operate from alternate premises
Supplier, customer, and client base intact	Supplier, customer, and client base disrupted, possibly for some time
Employee and well-being unaffected or managed through existing mechanisms	Need for dedicated services to support health and well-being in employees and families

how the business and its employees will act to ensure that core business objectives can be pursued should disruption occur. To achieve these objectives, planning must accommodate the following (Elliott et al., 2002; Business Continuity Institute, 2002):

- Understand the critical processes required to ensure the supply of goods and/or services to customers, provide income for the business, and maintain employment and consider events elsewhere that could affect supply and consumption (even if not directly affected by a disaster).
- Identify potential risks to the business in the context of its geographical position (e.g., the hazard-scape in which its core activities occur).
- Conduct a business impact analysis to assess risks associated with hazard consequences and how they will be distributed over key operations and activities.
- Consider strategies and options available to mitigate identified risks to the business (e.g., increasing the amount of insurance to transfer the risk; improving the structure of the building to withstand severe weather or earthquakes; installing efficient back-up systems at remote locations, etc.).
- Identify the time frame in which this risk must be borne, how long it will take for full recovery to occur, and the implications this will have for operations and for staff and staffing.
- Draw up a business continuity plan that identifies the actions the business will take in the event of a disaster and allocate roles and responsibilities for implementation accordingly and over an appropriate time frame.
- Train staff and embed a culture of BCM within the business.

Putting Business Continuity Management into Practice

No two businesses are alike and each will experience a given disaster in different ways (e.g., as a result of differences in size, manufacturing versus service industries, supply and distribution needs, etc.). Consequently, BCM plans and activities must be tailored to the needs of each business. Each business must identify the implications of disruptions to its core business elements, what resources are available to it to assist in developing contingency and disaster recovery plans, what additional resources it would need to develop or acquire to support recovery, and develop contingencies to cater for not being able to use or acquire these resources. This may include considering social responsibilities, and developing mechanisms for integrating business continuity planning with societal level planning to support social and economic recovery. An important aspect of business preparedness is ensuring that the BCM includes a checklist of "who does what" in the event of a disaster and identifies practical strategies for cooperating with, for example, the emergency services, suppliers and customers, the utility companies, local authorities, insurance companies, other businesses in the area, and the communities in the area in which it operates.

To be carried out effectively, BCM requires an adequate allocation of resources, both financial and human. Consequently, managerial acceptance of risk and their commitment to BCM is essential to planning being initiated and developed to an appropriate state of readiness. The culture of the organization, particularly with regard to its approach to strategic change, plays a role in determining whether this need is identified and acted on. The development of an appropriate change and risk management culture is a challenging task, especially for small businesses. The beliefs of managers and other key players can influence this process.

Business commitment to disaster continuity planning can be constrained by managers overestimating existing capabilities and by ambiguity of responsibility (Folke et al., 2003; Gunderson et al., 1995; Shaw & Harrald, 2004). Issues regarding responsibility are particularly important. A precursor to effective BCM is having responsibility vested in a key figure who can direct and sustain the planning process (Shaw & Harrald, 2004). For small business, entities such as Chambers of Commerce and industry groups could help facilitate this process (e.g., through mentoring), particularly with regard to specialist areas such as planning, training, and developing capability.

Preparing a business to cope with and recover from a disaster involves developing a capability to manage disruption from events that in most cases have not occurred and that could present in a context of widespread societal disruption and devastation that is difficult to anticipate and comprehend. Yet, managers must confront this task armed primarily with experiences derived from their own business history and the performance of routine activities.

Organizational Culture, Learning, and Change

Promoting effective change in the direction of accepting the need for and implementing disaster business continuity capability requires understanding how managers think about future, non-routine events about which there is considerable uncertainty. This task is complicated by the fact that, over time, the conceptual frameworks or "mental models" (see Chapter 2) that inform managers' thinking and action become entrenched in routine activities. This can make it difficult for them to anticipate the kinds of non-routine contingencies that may have to be managed should a disaster occur (e.g., Paton & Wilson, 2001). To counter these problems, it is important that business representatives engage in discussion with experts (e.g., scientific and risk management agencies) in ways that challenge assumptions, help them anticipate currently unforeseen circumstances, and help them identify the steps that can be taken to plan for and mitigate hazard consequences that could adversely affect organizational survival. An important issue in this context is understanding how organizational cultural processes can constrain managerial and organizational action.

Several factors capable of preventing business representatives engaging in the kind of critical debate required to support business continuity have been identified. These include thinking that crises only happen to other organizations (see unrealistic optimism–Chapter 4) or that the organization is too big and powerful to be affected by a disaster (Mitroff & Anagnos, 2001). Such beliefs can prevent change, or render the planning and implementation of change a more challenging endeavor. Other constraints arise where power and authority are highly centralized (Gunderson et al. 1995; Harrison & Shirom, 1999). In highly centralized businesses, bureaucratic inertia and vested political interests can conspire to block change and, indeed, sow the seeds of future and more complex crises as a consequence (Gunderson et al., 1995). Other influences include the attitude that some government or other external agency will come forward to help, or that disasters should be accepted as an inevitable part of life for which there is no point in preparing. Businesses that assume that the community will automatically return to pre-existing conditions after the disaster are less likely to survive it (Alesch, Holly, Mittler & Nagy, 2002).

A second type of reaction is where the business tries to plan, but lacks appropriate experience to do so effectively. This can occur as a consequence of a failure to consider the non-routine nature of the challenges disasters present businesses (e.g., damage to premises, loss of supply chains) or because the organization has failed to learn lessons from previous events (Folke et al., 2003; Paton et al., 1998). In this scenario, whether disaster response leads to greater resilience or heightened vulnerability is determined more by chance than by sound planning and good judgment. In general, businesses that have sufficient structural flexibility are in a better position to respond effectively to crises (Alesch et al., 2001; Folke et al., 2003; Paton, 1997), but still need to develop the capacity to do so.

Folke et al. (2003) emphasized the fact that to increase resilience, some experience of failure, and an ability to learn from it, is often required. Failure provides valuable insights into areas where development is required. The idea that a business should plan for failure as well as success is a difficult concept to accept. However, "failing to plan to fail" is as important as "failing to plan to succeed" (Folke et al., 2003) as detriments to business continuity, particularly when anticipating the widespread loss and destruction that can accompany a large-scale disaster.

Not only must the organization develop a culture that embraces a need to learn to live with risk and uncertainty, it must develop strategies to learn from the unexpected disturbances and failures that arise over time (both to itself and to other businesses in the same industry sector). Recognition of the importance of institutional learning leads to a third strategy, one capable of actively contributing to business capacity to cope, adapt, and recover.

According to Folke et al. (2003), this involves several activities. First, it requires the memory of prior crises, with personal experience of a disaster or knowledge of a disaster in a neighboring or similar business being potent motivational factors (Dahlhamer & D'Souza 1995), with the lessons learned being incorporated into institutional memory and part of the organizational culture. Second, it requires a commitment to learn from experiences in ways that develop future capability. Finally, these lessons must be encapsulated in new policies and procedures that facilitate organizational adaptability. The effectiveness of this institutional learning approach can be enhanced by creating small-scale, controlled disturbances (e.g., realistic simulations or exercises) to facilitate the learning process and challenge complacency (Coutu, 2002; Folke et al., 2003; Paton & Jackson, 2002; Paton & Wilson, 2001). One of the outcomes of this process is the identification of the competencies and capabilities required of the staff that will be responsible for implementing the plan during a disaster. That is, it provides an input into training needs analysis and design.

TRAINING

The environment within which businesses will be required to operate during and after a disaster will differ substantially from that in which routine business activity is undertaken. If the benefits of BCM planning are to be fully realized and can be implemented in a timely and effective manner, staff capable of applying them in a context defined by a need to confront challenging circumstances must be developed. This can be accomplished by including training activities in the planning process.

The effectiveness of the business response to disaster is a function of staff having the competencies (e.g., information management) and capabilities (e.g., stress resilience) required to put plans into action under atypical and challenging circumstances (Grant, 1996; Paton, 1997a). To do so, business preparation planning should include training needs analysis conducted explicitly to identify the consequences likely to be encountered during and after a disaster and identifying the competencies required to manage them. Significant differences between routine and post-disaster environments create novel and highly challenging demands for managers. Training is thus required to enhance their response capability (Paton & Jackson, 2002). Training should cover, for example:

- risk assessment and its implications for staff well-being and operational continuity;
- developing a managerial style suited to identifying and planning to meet staff and business needs under crisis conditions that may persist for a considerable period of time;
- adopting a decision style suited to operating under conditions of uncertainty;
- familiarization with response plans and procedures
- using problem-solving skills to adapt plans and procedures to the contingencies presented by a disaster;
- operating under devolved authority and planning for management succession (into crisis roles and from crisis back into routine operations);
- communicating and working with people with diverse backgrounds and abilities;
- reconciling staff and business recovery needs (over time), and
- staff monitoring and managing the return to work process.

Given the infrequent nature of hazard events, training that includes exercises and simulations can provide opportunities to gain some idea of what the operating environment would be like. Simulations afford opportunities for staff to develop the technical and managerial skills that will be required to respond to hazard consequences, practice their use under adverse circumstances, receive feedback on their performance, increase awareness of stress reactions, and rehearse strategies to minimize negative reactions (Flin, 1996; Paton & Jackson, 2002; Rosenthal & Sheiniuk, 1993). To enhance response effectiveness, planning should inform the development of crisis management procedures.

Crisis management systems are required to cover, for example, delegation of authority; allocation of crisis response tasks, roles, and responsibilities and the development of appropriate management procedures; identifying and allocating resources necessary to deal with the crisis, information management, communication and decision management, and liaison mechanisms. Flexibility in these systems is important. They will be required to deal not only with the uncharacteristic demands of the crisis, but also atypical demands emanating from dealing with unexpected emergent tasks, dealing with unfamiliar people and roles, and frequent staff reassignment (Folke et al., 2003; Paton, 1997). Communication systems, designed to meet the needs of diverse stakeholder and response groups, are required for information access and analysis; defining priority problems; guiding emergency resource needs and allocation; coordinating activities; providing information to managers, staff, and the media; and for monitoring staff and business needs (Bent, 1995; Doepal, 1991; Paton, 1997), possibly over a period of several months.

Continuity planning and preparedness play important roles not only in facilitating the continued existence of the businesses in areas affected by disaster but also in contributing to person, household, community, and economic well-being in areas affected by disaster. It is important to acknowledge that people, communities, and businesses play complementary roles in disaster response and recovery. However, responsibility for developing this capacity also rests with people, communities, and businesses.

Analyses of people's response to recent events such as the Black Saturday bushfires in Australia and the Christchurch, New Zealand earthquake have highlighted the important role that relationships between communities and agencies and between communities and businesses play in effective recovery. With this experience has come the recognition, from both citizen and business perspectives, of the need to develop these relationships in advance. This means expanding the scope of preparedness programs to include encouraging the development of community and business preparedness and the awareness amongst all stakeholders (business owners, employees, citizens, government, and risk management agencies, etc.) that this kind of complementary preparedness lays the foundations for faster and more effective recovery through sustaining economic vitality and people's livelihoods. From a business perspective, adding a social responsibility element to business disaster continuity planning can facilitate attaining this outcome by helping businesses recognize the interdependencies that exist between employee (citizen) and business preparedness. This builds on the individual/household and community preparedness discussed earlier to develop a comprehensive societal capacity to cope with, adapt to, recover from, and learn from disaster experience.

Chapter 10

FUTURE ISSUES IN HAZARD PREPAREDNESS: ENGAGING PEOPLE, SCIENCE, PRACTICE

Tell me, and I will forget,
Show me, and I may remember,
Involve me, and I will understand.
Confucius

Worldwide, members of many societies live with the possibility of experiencing adverse impacts from natural processes (e.g., volcanic, wildfire, storm, flooding, tsunami, and seismic) whose activity and actions can be hazardous when they interact with people and what people value and/or rely on to sustain everyday life (see Chapter 1). Recognition of this susceptibility has prompted the active pursuit of strategies to manage the associated risk. One important risk management goal in this context is encouraging people to prepare for hazard events. Engaging in activities such as critically discussing hazard issues with others; collaborating with others to develop household, neighborhood, and community emergency plans and capabilities; securing the physical integrity of the home and household items; and storing food and water are examples of actions that contribute to people's ability to anticipate what they might have to contend with and to developing their capacity to cope with, adapt to, recover from, and learn from hazard events.

The benefits of preparing in these ways has become increasingly evident from their being identified by people who have experienced disasters as knowledge, resources and activities that could make crucial contributions to their response and recovery (Mamula-Seadon, Selway, & Paton, 2012; McLennan & Elliot, 2012; Paton & Tang, 2009; Paton, 2012). Thus when put into practice when disaster strikes, preparedness functions to increase community and societal resilience (Paton & Johnston, 2006). However, despite the evident advantages that being prepared confer on people and communities, research has consistently found that individual, community, and business preparedness levels tend to be low. This book examined both why this is so and what can be done to expedite the development of sustained preparedness, at household, community, and societal levels.

The Confucian quote that opens this chapter encapsulates both one of the main reasons why many contemporary attempts to encourage preparedness have fallen short of expectations and how future approaches can be rendered more effective. The Confucian maxim, "Tell me, and I will forget" neatly sums up research findings that indicate that simply giving people information about hazards and risk, irrespective of its quality, is not necessarily sufficient to motivate action. This does not mean that knowing one's risk and having access to high quality hazard and risk information are unimportant.

Knowing one's risk is vital. Without some level of risk perception and acceptance, people are very unlikely to do anything. However, knowing one's risk and knowing what to do, how to do it, and being able to do it are not the same. These capabilities need to be developed. Similarly, good information plays an important part in how hazard preparedness capability develops and it remains vital that risk communication strategies know how best to present information to people (e.g., Mileti, Nathe, Gori, & Lemersal, 2004). But information on its own, without considering how people interpret and then use it, will not advance preparedness as much as might be expected. Rather, as Confucius, some 2,500 years ago, pointed out, if the goal is to promote understanding (and action), then risk management programs must not only inform people, but also engage them in ways that facilitate the understanding and that develop their capacity to act.

The contents of this book have focused primarily on identifying how to involve or engage people. Engagement was discussed from several perspectives: how people engage with or relate to their environment (which was described in terms of how people can learn to co-exist with the potentially hazardous aspects of its environment); how people engage with each other; and how people engage, both directly and indirectly, with scientists, risk management professionals, businesses, and the media. The common denominator is that, conceptually, all these facets of engagement occur in the context of risk management. In particular, this defines the context in which the stakeholders, directly and indirectly, interact and in which (filtered and interpreted) information is disseminated to people and groups who differ with regard to expertise, needs, goals, expectations, time, and capability. When risk management is conceptualized in this way, it highlights the fact that stakeholders differ substantially with regard to their views on preparedness and how they interpret and use information provided with the objective of encouraging them to prepare.

If citizens took scientific and risk management information at face value, it could be anticipated that they would accept and act on expert information and recommendations. However, citizens rarely take information at face value. Rather, people actively engage in interpreting and making sense of information and, indeed, their (hazardous) circumstances. It is the outcome of this sense-making process that guides what people do, or don't do.

How engagement influences risk interpretation and action was explored from three general perspectives. The first discussed how community engagement became a requisite component of comprehensive risk management as a consequence of the fact that when they enter the risk management process, citizens tend to make sense of their hazardous circumstances in ways that differ quite substantially from their scientist and risk expert counterparts (see Chapter 2). Reconciling these perspectives in ways that facilitate the ability of all stakeholders to share responsibility for risk management and contribute to it in complementary ways can be facilitated by ensuring that people, science, and practice engage in ways that can achieve this kind of complementarity (see also Chapters 5–7). The second perspective was in relation to what preparedness is (see Chapters 3). The third was concerned with theories developed to capture how intra-individual, social, and relationship variables interact to influence how people access and use information to interpret their circumstances and their risk and make decisions about how to manage their risk (see Chapters 4–7). The role of community engagement was particularly evident in the Protective Action Decision Model and Community Engagement theories discussed in Chapters 5 and 7. Given that this book is about preparedness, it is appropriate to commence the concluding chapter by reiterating the need for more critical assessment of what is trying to be achieved: what is comprehensive preparedness?

WHAT CONSTITUTES COMPREHENSIVE AND EFFECTIVE PREPARATION?

The object of preparedness research is to account for differences in levels of preparedness. The existence of different functional categories (see Chapter 3) means that preparedness research is faced with the challenge of explaining both differences in the levels of the adoption of discrete functional categories and explaining differences in levels of preparedness within each functional category. A tendency to have conceived of preparedness as a singular phenomenon, rather than as a suite of distinct, but related (within the recovery process), functional categories, may have obscured the need to explain the former.

This book argued that effective and comprehensive natural hazard preparedness is a multi-functional and multilevel phenomenon. The latter describes how the salience of different aspects of preparedness changes with the level of analysis. At the household level, preparedness includes securing the structural integrity of the house and property (e.g., securing walls to foundations, creating a defensible space) and securing internal fittings and fixtures. At the individual and family level, preparedness involves taking steps to understand personal risk and developing knowledge, beliefs, competencies, and the behaviors required to secure the resources needed to be self-reliant in relation to the hazard consequences they could encounter. It also encompasses the development of household plans that cover the needs of the family and its members under all foreseeable circumstances (e.g., events occurring in evenings or at weekends when all members are at home, events occurring during weekdays when family members may be separated as a result of attending work and school, events requiring prolonged relocation, the needs of children at different ages, and so on) that can be enacted promptly should the need arise and that can be sustained over an appropriate period of time (e.g., permanent or prolonged relocation, loss of or disruption to livelihood).

At the community level, preparedness is about developing both neighborhood and community knowledge (e.g., inventories of the skills and capabilities available within a community) in conjunction with the capacity to use local knowledge and resources to respond to issues collectively in ways that meet the needs of people within a neighborhood or community (see Chapter 7). At the community/societal level, effective preparedness also encompasses the development of relationships between people and civic agencies, local government departments, and businesses. While not providing a comprehensive account of preparedness at the societal level, the key issues of what organizations (especially businesses, but also other societal organizations) were addressed in a discussion of business continuity and recovery planning (see Chapter 9).

If society is to fully realize the potential return on investment that can accrue from the resources invested in comprehensive risk management planning (i.e., to contribute to overall societal resilience), it is important that this planning adopts a correspondingly comprehensive approach. The interrelationships that exist between levels of preparedness mean that the development of preparedness at only one level will result in response and recovery falling short of what could be achieved.

These examples call into question whether preparedness is a phenomenon that can be treated as a one homogenous entity in preparedness research. Yet, this is what tends to happen in preparedness research. Preparedness measures are generally treated collectively as a single dependent variable with little regard to the items people check in questionnaires (see Chapter 3). What is needed is a more critical analysis of the relationship between each functional category and its antecedents. In addition to its theoretical implications, the outcomes of such analyses have practical ramifications for the design and delivery of risk communication and community engagement strategies.

Despite their all playing a role in assisting response and recovery, preparedness measures (e.g., structural, planning, and relationship, etc. elements—see Chapter 3) can be differentiated with regard to the respective functional roles they play in helping people and communities cope, adapt, and recover as they progressively negotiate the impact, response, and recovery phases of disaster. They perform different functions and their relative salience changes over the course of the recovery process (see Chapter 3).

Evidence for functional differences in preparedness emerged from two sources. The first was from empirical research (see Chapter 3) and the second from studies of what community members report helping them cope, adapt, and recover following their experience of significant hazard events such as the 2004 Indian Ocean tsunami (Paton & Tang, 2009) and the 2011 Christchurch, New Zealand earthquake (Mamula-Seadon et al., 2012; Paton, 2012). Analyses of people's accounts of their response and recovery experiences also highlighted the need to include an additional functional category, psychological preparedness, in inventories of preparedness measures. The fact that functional categories could be differentiated with regard to the competencies and resources required for their development (see Chapter 3) calls for a more searching analysis of the antecedents or predictors of preparedness. While they have not been applied to each functional area, existing theories provide rich veins of predictor variables ready to be mined and applied to the task of identifying functional antecedents (see Chapters 4–7).

ACCOUNTING FOR DIFFERENCES IN THE LEVELS OF FUNCTIONAL PREPAREDNESS

The discussion of the interpretive and meaning making processes that need to be accommodated in risk communication commenced with intrapersonal processes (see Chapters 2, 3, and 4), progressed through the social-cognitive and social processes (see Chapter 5 and 6) and culminated in the discussion of how people's social and societal relationships (see Chapter 7) influenced preparedness. These chapters identified how competencies and social processes at personal, social, and societal levels of analysis (e.g., fatalism, collective efficacy, trust, etc.) interacted to influence how people interpret their hazardous circumstances and made decisions, individually and collectively, about how to manage the risk natural hazards pose for them. The contents of Chapters 4 to 7 should be regarded as offering different, but complementary, insights into preparedness but not necessarily competing ways of explaining preparedness.

Though they differ with regard to the variables used and how these variables are posited to influence preparedness, collectively these theories offer diverse insights into the complex process that is hazard preparedness. Jointly, these theories identify variables that can be used to start this process.

As a starting point for this endeavor, several of the key variables and processes discussed in this book are summarized in Figure 10.1. Figure 10.1 describes the predictor variables and, in the right-hand column, the need for them to predict each of several preparedness categories. Figure 10.1 also includes a summary of the PADM theory (see Chapter 5) to illustrate how the search for information could be included in comprehensive model.

At the same time, work on identifying new variables and relationships needs to continue in order to realize the goal of developing comprehensive understanding of preparedness, its predictors, and the means of mobilizing these predictors in risk management programs. For example, this book has only briefly touched on predictor variables that could be sourced from research into family and family role relationships. Yet, there is evidence to suggest this might be a fruitful line of inquiry.

Gender role relationships can influence family decision making, as can the degree of collective family support for or conflict about household preparedness (Goodman & Cottrell, 2012; Paton & Buergelt, 2012). Family and role relationships have been implicated as being able to influence how family members deal with adverse and challenging circumstances (Paton & Norris, in press). Paton and Norris also discussed how the work-family interface can influence this process. This provides another example of how family and organizational (business) preparedness can play complementary roles in the development of comprehensive community preparedness. Consequently, some of these family issues were included in Figure 10.1 to illustrate their potential role in the preparedness process. Another issue that has received limited attention and which deserves a more prominent position in preparedness research concerns the developmental nature of preparedness.

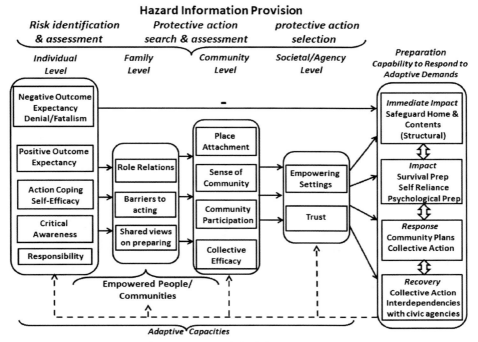

Figure 10.1. Composite model of the preparedness process.

PREPAREDNESS AS A DEVELOPMENTAL PROCESS

Preparedness often tends to be viewed as an all-or-none type of process. The review of preparedness theories in Chapter 5, however, discussed how the application of the Trans Theoretical Model (or Stages of Change) theory prompts thinking about preparedness as a developmental process in which people develop progressively more comprehensive levels of preparedness over time (and may even regress to earlier or less well-developed levels of preparedness over time). Some of the studies reviewed in Chapter 5 identified how the salience of predictor variables changed depending on the stage of preparedness people were embarking on. The systematic investigation of the relationship between predictor variable and stage in the development of preparedness is another area that needs to be included in future research agenda. This section considers what can be done to start this process and to offer some insights into what this might mean for intervention processes.

Previous work in this area did not consider the fact that some people decide not to prepare (see Chapters 3–5). Consequently, this section discusses strategies that could be pressed into use to help switch people from a "not-preparing" to a "preparing" orientation, how to get people started, how to facilitate the progressive development of people's preparedness, and how to sustain high levels of preparedness. Conceptualizing preparedness as a developmental process highlights the need for agencies and communities to interact over time. As such, it provides another example of how sustained preparedness is a function of how people, science, and practice engage in the context of the development and implementation of risk management strategies.

Strategies for Those Who Decide Not to Act

Dispositional characteristics and beliefs such as fatalism and negative outcome expectancy can result in people not taking action (see Chapters 4 and 7). Strategies that can help circumvent this tendency include presenting people with information about how hazard damage is selective and related to factors such as building design or environmental management (e.g., creating a defensible space to mitigate wildfire risk). The objective of the process at this level is to help people differentiate uncontrollable hazard events from their more controllable consequences and to demonstrate how this can be accomplished. Another strategy discussed involved inviting people to identify what they might recommend to help more vulnerable members (e.g., residents in a home for the elderly) of a community to prepare. These strategies can help shift people's thinking to a point where they are more open to considering preparing as an option. However, getting people to start to act requires other approaches.

Getting People Started

Often the starting point for preparedness programs is advising people of their susceptibility to experiencing adverse hazard consequences and presenting people with comprehensive accounts of the sources of their susceptibility and what they should be doing to prepare. This can be counterproductive for those just stepping onto the first rung of the preparedness ladder. As discussed earlier (see Chapter 4), presenting people who have yet to prepare with information about their susceptibility to experiencing highly damaging events can stimulate levels of anxiety or denial that can prevent people taking any steps to prepare. Similarly, for those just starting to prepare, presenting comprehensive lists of preparedness items and activities can overwhelm people and, by attracting their attention to the costs of taking action, reduce the likelihood that they will take the first step. These problems can be compounded if information about hazard behaviors such as intensity and magnitude (see Chapter 1) are provided to people at this stage in the developmental process. Alternative approaches to motivating people to prepare are required.

One approach involves getting people to first personalize their risk. This can be done by, for example, inviting people to think about the personal consequences a hazard event could have for them (e.g., asking them how it could affect them, their family, and their livelihood, etc.). Other approaches include first asking people to generate ideas about why preparing might be valuable or presenting information to people sourced not from experts but from people in similar communities or circumstances (e.g., their views on risk, what happened and, preferably, what worked from them when they experienced a hazard event) to those being targeted by a risk communication or outreach program. These approaches can not only help people personalize risk but intrinsically motivate people to develop their interest in taking things further. Intrinsically motivated people are more likely to take responsibility to develop their preparedness and move to the next step.

Encouraging Preparedness

Strategies designed to personalize issues and motivate action also need to be cognizant of cognitive biases, such as normalization bias, the gambler's fallacy, unrealistic optimism, and risk homeostasis. The adverse effect these may have on preparedness can be countered more effectively once people are motivated to act and they start their preparedness. These biases can be countered by, for example, encouraging people to discuss hazard with others to draw attention to what people have and have not done and, through facilitating discussions of the benefits of preparedness and how to achieve it, create the conditions in which preparedness norms are more likely to develop.

At this stage, when people are (more) intrinsically motivated to prepare, information about risk and the nature of the hazard consequences they could experience (e.g., regarding hazard characteristics and behavior–see Chapter 1) is more likely to prompt further preparedness. However, the way this information is presented is important. In particular, it is important to provide information in coherent, small chunks. For example, when discussing earthquake or wildfire preparedness, present information on one hazard (e.g., ground shaking or ember attack) at a time, explain how a hazard creates losses (e.g., how ground shaking can lead to building collapse or how embers ignite buildings), and elucidate how preparedness measures can prevent or minimize losses (e.g., securing a house to its foundations limits damage from shaking or how a defensible space limits ember attack). Thus, information is more likely to be understood when presented in coherent, meaningful chunks that explain the function of preparedness measures in the context of explaining the hazard and its consequences. If possible, this information should be provided in social contexts (e.g., community group meetings) and/or encourage people to discuss issues provided through risk communication with others.

Encouraging risk management discussion to occur in social contexts is important in several respects. For example, it accommodates the fact that when faced with complex and uncertain events, people's perception of risk and how they might manage this risk are influenced by information from others who share their interests and values (see Chapter 7). Engaging people in community contexts thus provides opportunities for people to develop their risk beliefs, ask questions and seek input, correct misconceptions, and so on and to relate these issues to their own circumstances.

Engaging people in this way can also contribute to psychological preparedness through, for example, increasing people's knowledge and sense of control and by creating a context in which social support (e.g., informational, emotional, tangible, esteem, belonging, etc.) becomes more readily available. Social support can not only help people through the process of preparing but also provide an important recovery resource in the event of a disaster. Getting people actively involved in these activities can provide opportunities for them to apply and/or develop core competencies such as collective efficacy. Working with risk management agencies in a community-centered way can also contribute to empowering community members. Other competencies that can facilitate this process include prioritizing, planning, and implementation skills. As people develop their knowledge and preparedness, it becomes necessary to develop strategies to sustain this momentum and so sustain preparedness.

Sustaining Preparedness

As people become more involved and committed to preparing, this momentum can be sustained in several ways. For example, it can include providing (group-based if possible) activities that include demonstrations and property assessments where well and poorly prepared properties are visited and discussion around, for example, what works, what needs changing, how, and so on can be used. Several activities, such as search conferences, design meetings, and workshops, can be organized to help advance community thinking, planning, and action. The development of community liaison committees, with elected community leaders, can facilitate the capacity of a community to formulate its ideas and be able to represent them to risk management, government agencies, and businesses in ways that further community preparedness. Such activities, as more structured mechanisms for working with agencies, can contribute to empowerment and trust (see Chapter 7).

One issue that defining preparedness as a developmental process highlights is the practical challenge of trying to engage with neighborhoods whose members could be at different stages of development or who need different developmental inputs to help them transition to more comprehensive levels of preparedness. It is possible that focusing on this higher level, the sustained preparedness level, may offer (cost-effective) approaches to assisting community members whose preparedness is less well advanced. This could be accomplished by using collaborative learning and peer tutoring techniques. This involves those more advanced and knowledgeable community members acting as tutors, mentors, or advocates for those at lower levels of preparedness or as general sources of advice and information for other community members. This would allow risk management agencies to focus on those with an existing commitment to preparedness and focusing on how to facilitate peer or collaborative learning and providing training for peer tutoring.

Keeping preparedness activities within a neighborhood or community has several advantages. Not only does it enable the development of strategies more likely to accommodate local issues, needs, and goals, it also represents a more appropriate context for the development and maintenance of core community characteristics and competencies such as being actively engaged in community life and self- and collective efficacy. Those already committed, as was evident in Chapter 7, will possess these competencies and working with others to identify and resolve local risk management issues and implementing solutions will foster their development in others. It is also, at this point, worth reflecting on how these competencies develop in everyday life and to consider what this means for community-based risk management. This refers to considering the origins and levels of the variables included in theories of preparedness.

INTEGRATING RISK MANAGEMENT AND COMMUNITY DEVELOPMENT

The theories discussed in this book indirectly offer novel insights into the cost-effective development of preparedness strategies. To understand why it is necessary to briefly revisit the fact that the predictors of preparedness identified earlier (see Chapters 4–7) did not result from people having special training in risk management. The competencies and relationships people use to make sense of information and make their decisions derive from their more mainstream life experiences.

For example, people who bring experience of successfully negotiating challenge and change in everyday life will bring with them a greater sense of self-efficacy to their preparedness activities (which influence things like the range of issues people include in their plans, the persistence in activities). In contrast, those who may approach preparedness with a history of having been thwarted in their attempts to achieve life goals or deal with challenging tasks in life are more likely to enter the risk management context with a stronger sense of learned helplessness that would serve to reduce their propensity to prepare. Similarly, at the point when they enter into a risk management process, people can be differentiated with regard to their experience of collaborating with others (e.g., levels of community participation) and with regard to the degree to which such collaborative endeavors have resulted in people being engaged in activities that require, for example, the analysis of challenging circumstances and engaging with others to solve problems in ways that facilitate their attaining successful outcomes within their community. The latter facet of social experience is captured in the collective efficacy concept. All of these have important roles to play when people are confronted by disaster (Mamula-Seadon et al., 2012; McLennan & Elliot, 2012; Paton, 2012; Frandsen et al., 2012).

Recognizing the role of cumulative everyday life experience in developing the competencies that predict natural hazard preparedness opens up opportunities for developing levels of these antecedents by integrating risk management strategies with those that focus on, for example, community development (Anckermann et al., 2005; Paton & Jang, 2011). Community development activities provide people with the kinds of experiences that increase their level of engagement with others and provide them with opportunities to develop competencies such as collective efficacy.

Thus engaging in these community-oriented activities will increase people's ability to interpret and use risk management information provided through risk communication and community engagement programs. Similarly, workplaces, religious institutions, schools, and tertiary institutions could be used as contexts for integrating risk communication and competency development, and do so in ways that increase learning opportunities in community contexts. How this might be accomplished is another area for future research and one that can focus on how science and practice can profitably engage together.

Taken together, the issues canvassed in this book suggest that, after some 2500 years, risk management planning is in a better position to realize the benefits of the Confucian maxim that if people are involved, they will understand. If they understand, it becomes easier to encourage people to act. A combination of understanding and action places people and communities in a better position to develop the individual and collective competencies and the social relationships and processes required to facilitate their ability to co-exist with hazardous consequences that arise from living with unremitting geological and environmental forces. The importance of developing the individual and collective capabilities of people, communities, civic agencies, and businesses that underpin a capacity to co-exist with hazards can be traced to the ever-increasing exposure of people and societies to hazardous circumstances.

HAZARDOUS FUTURES

The quotation that introduced this book (see Chapter 1) highlighted the precarious position humankind faced from exposure to unpredictable geological and environmental forces. These forces of nature are not going to disappear and may indeed occur with greater frequency and with greater destructive potential in the future. It can thus be realistically anticipated that people and societies will have to contend with ever more threatening and complex hazard-scapes, with these changes emanating from both social and environmental change over coming years and decades.

Threats to people and what people value is increasing as a result of factors such as population growth and development pressures that see people encroaching ever more on areas that increase their exposure to the consequences of hazard activity (e.g., increased demand for affordable housing increasing building and development in flood prone areas, growth of the peri-urban fringe increasing exposure to wildfire hazards, development in coastal environments increasing exposure to storm and tsunami hazards). Thus, certain choices being made by people and societies that influence their relationship with the natural environment act to increase the likelihood of personal, community, and societal exposure to hazard events. If possible, this kind of exposure should be avoided (see Chapter 2). But if this is not an option (e.g., where extensive development has already occurred), societal development should be accompanied by the inclusion of strategies that progressively and sustainably facilitate people's ability to co-exist with the potential exposure to natural hazards that their lifestyle, economic, development, and growth choices have engendered. A need for efforts to facilitate a capacity of co-existence will also arise from changes to the hazard-scapes in which societies are situated.

From an environmental change perspective, the most obvious, and enduring, influence on people's future exposure to hazardous circumstances will occur as a result of climate change. Risk from this quarter will occur directly as a result of, for example, the increasing incidence of climate-related hazards such as storms and hurricanes (cyclones and typhoons) and increased incidence of wildfire hazards. Climate change may also have indirect influences on hazard-scapes.

For example, McGuire (2012), based on an analysis of events following the last ice age, draws attention to an interesting relationship between temperature-related loss of ice sheets and earthquake, tsunami, and volcanic activity. The loss of ice sheets could result in the earth's crust bouncing back from millennia of being depressed by the weight of ice and, according to McGuire, increase the risk of hazard events. Reports of faster than anticipated declines in Arctic ice sheet thickness in late 2012 add to the need to consider the potential significance of this issue, particularly if this risk occurs in areas that may not have previously afforded seismic hazards a high priority in their hazard-scapes.

As the quote that opened this book pointed out, nature can and will continue to impose change on people and will continue to do so suddenly. However, people, communities, societies, and businesses can make choices about how they will co-exist with natural and environmental hazards. They can, through various means, advance their understanding of these processes, mitigate potential consequences, increase the degree of notice of impending activity available to them, and prepare in ways that allow them to take more control over the degree of change impose by natural forces on them.

While a need for more research into all these facets of risk management remains, this book has provided a comprehensive account of what can currently be done to increase the capacity of people, communities, societies, and businesses to anticipate what they may have to contend with (hazard consequences), cope with, adapt to, recover from the hazard consequences they experience, and to learn from experiences in ways that contribute to the development of future societal resilience and adaptive capacity. All that remains is for people, communities, agencies, and businesses to find the will and the resolve to put these ideas into practice.

REFERENCES

Abraham, C., Conner, M., Jones, F., & O'Connor, D. (2008). *Health psychology.* London: Hodder Education.

Abramson, L. Y., Seligman, M. E. P., & Teasdale, J. A. (1978). Learned helplessness in humans: Critique and reformulation. *Journal of Abnormal Psychology, 87,* 49–74.

Ahn, W.-K., & Bailenson, J. (1996). Causal attribution as a search for underlying mechanisms: An explanation of the conjunction fallacy and the discounting principle. *Cognitive Psychology, 31,* 82–123.

Aitken, C., Chapman, R. B., & McClure, J. (2011). Climate change, powerlessness and the commons dilemma: Assessing New Zealanders' preparedness to act. *Global Environmental Change, 21,* 752–760.

Alesch, D. J., Holly, J. N., Mittler, E., & Nagy, R. (2002). *Organizations at risk: What happens when small businesses and not-for-profits encounter natural disasters* [http://www.riskinstitute.org/ptr_item.asp?cat_id=1&item_id=1028], accessed 14th February 2003.

Alesina, A., & La Ferrara, E. (2000). Participation in heterogeneous communities. *Quarterly Journal of Economics, 115*(3), 847–904.

Alexander, D. (1993). *Natural hazards.* New York: Chapman and Hall.

Anckermann, S. M., Dominguez, N., Sotol, F., Kjaerulf, P., Berliner, & Mikkelsen, E. N. (2005). Psycho-social support to large numbers of traumatized people in post-conflict societies: An approach to community development in Guatemala. *Journal of Community & Applied Social Psychology, 15,* 136–152.

Anderson, J. C., & Gerbing, D. W. (1988). Structural equation modeling in practice: A review and recommended two-step approach. *Psychological Bulletin, 103*(3), 411–423.

Anderson-Berry, L. J. (2003). Community vulnerability to tropical cyclones: Cairns, 1996–2000. *Natural Hazards, 30*(2), 209–232.

Andreasen, A. R. (1995). *Marketing social change: Changing behaviour to promote health, social development, and the environment.* San Francisco: Jossey-Bass.

Armitage, C. J., & Christian, J. (2003). From attitudes to behaviour: Basic and applied research on the theory of planned behaviour. *Current Psychology, 22*(3), 187–195.

Armitage, C. J., & Conner, M. (2001). Efficacy of the theory of planned behaviour: A meta-analytic review. *British Journal of Social Psychology, 40,* 471–199.

Atman, C. J., Bostrom, A., Fischhoff, B., & Morgan, M. G. (1994). Designing risk communications: Completing and correcting mental models of hazardous processes, part I. *Risk Analysis, 14*(5), 779–788.

Atwood, L. E., & Mdor, A. M. (1998). Exploring the "Cry Wolf" hypothesis. *International Journal of Mass Emergencies and Disasters, 16,* 279–302.

Ajzen, I. (1985). From intentions to actions: A theory of planned behavior. In J. Kuhl & J. Beckmann (Eds.), *Action control: From cognition to behavior* (pp. 67–69). Berlin, Heidelberg, New York: Springer-Verlag.

Ajzen, I. (1991). The theory of planned behavior. *Organizational Behavior and Human Decision Processes, 50,* 179–211.

Ajzen, I., & Fishbein, M. (1980). *Understanding attitudes and predicitng social behavior.* Englewood Cliffs, NJ: Prentice Hall.

Bagozzi, R. P. (1992). The self-regulation of attitudes, intentions and behaviour. *Social Psychology Quarterly, 55,* 178–204.

Bagozzi, R. P., & Dabholar, P. A. (2000). Discursive psychology: An alternative conceptual foundation to the means-end chain theory. *Psychology and Marketing, 17,* 535–586.

Bajek, R., Matsuda, Y., & Okada, N. (2008). Japan's Jishu-bosai-soshiki community activities: Analysis of its role in participatory community disaster risk management. *Natural Hazards, 44,* 281–292.

Ballantyne, M., Paton, D., Johnston, D., Kozuch, M., & Daly, M. (2000). *Information on volcanic and earthquake hazards: The impact on awareness and preparation.* (GNS Science Report 2000/2). Lower Hut, New Zealand: Institute of Geological and Nuclear Sciences.

Bandura, A. (1977). Self-efficacy: Toward a unifying theory of behavioral change. *Psychological Review, 84*(2), 191–215.

Bandura, A. (1988). Organisational applications of social cognitive theory. *Australian Journal of Management, 13*(2), 275–302.

Bandura, A. (1997). *Self-efficacy and agency of change.* New York: Raven Press.

Barberi, F., Brondi, F., Carapezza, M. L., Cavarra, L., & Murgia, C. (2003). Earthen barriers to control lava flows in the 2001 eruption of Mt. Etna. *Journal of Volcanology and Geothermal Research, 123,* 231–243.

Barlow, D. H. (2002). *Anxiety and its disorders* (2nd ed.). New York: Guilford Press.

Barnett, D. J., Balicer, R. D., Blodgett, D. W., Everly, G. S., Omer, S. B., Parker, C. L., et al. (2005). Applying risk perception theory to public health workforce preparedness training. *Journal of Public Health Management Practice, November (Suppl.),* S33–S37.

Barnett, J., & Breakwell, G. M. (2001). Risk Perception and experience: Hazard personality profiles and individual differences. *Risk Analysis, 21*(1), 171–178.

Baron, J. (2000). *Thinking and deciding* (3rd ed.) Cambridge: Cambridge University Press.

Barton, A. H., & Lazarsfeld, P. F. (1956). *Some functions of qualitative analysis in social research.* New York: Columbia University Press.

Basili, M. (2006). A rational decision rule with extreme events. *Risk Analysis, 26*(6), 1721–1728.

Baumann, D. D., & Sims, J. H. (1978). Flood insurance: Some determinants of adoption. *Economic Geography, 54,* 189–196.

Baxter, J., & Eyles, J. (1999). The utility of in-depth interviews for studying the meaning of environmental risk. *Professional Geographer, 51*(2), 307–320.

Beatley, T. (1990). *Managing reconstruction along the South Carolina Coast: Preliminary observations on the implementation of the Beachfront Management Act.* Boulder, CO: University of Colorado, Natural Hazards Research and Applications Center.

Becker, M. H. (1974). *The health belief model and personal health behaviour.* Thorofare, NJ: Slack.

Bechara, A., Damasio, H., Tranel, D., & Damasio, A. R. (1997). Deciding advantageously before knowing the advantageous strategy. *Science, 275*(5304), 1293–1295.

Beierle, T. C. (2002). The quality of stakeholder-based decisions. *Risk Analysis, 22*(4), 739–749.

Bell, C., & Newby, H. (1971). *Community studies: An introduction to the sociology of the local community.* London: Allen and Unwin.

Bennett, P., & Murphy, S. (1997). *Psychology and health promotion.* Buckingham: Open University Press.

Bent, D. (1995). Minimising business interruption: The case for business continuance planning. In A.G. Hull & R. Coory (Eds.), *Proceedings of the natural hazards management workshop 1995* (pp. 15–20). Lower Hutt, New Zealand: Institute of Geological and Nuclear Sciences.

Berger, P. L., & Luckmann, T. (1967). *The social construction of reality.* Hammondsworth: Penguin.

Berkman, L. F. (1995). The role of social relations in health promotion. *Psychosomatic Medicine, 57*(3), 245–254.

Berrill, J. B. (1997). Seismic liquefaction and lifelines. In Christchurch Engineering Lifelines Group (Eds.), *A multidisciplinary approach to the vulnerability of lifelines to natural hazards.* University of Canterbury, Canterbury: Centre for Advanced Engineering.

Bhandari, R., Okada, N., Yokomatsu, & Ikeo, H. (2010). Analyzing urban ritual with reference to development of social capital for disaster resilience: A case study of Kishiwada. Proceedings of IEEE International Conference on Systems, Man and Cybernetics, pp. 3477–3482.

Bishop, B., Paton, D., Syme, G., & Nancarrow, B. (2000). Coping with environmental degradation: Salination as a community stressor. *Network, 12,* 1–15.

Bloom, P. N., & Novelli, W. D. (1981). Problems and challenges in social marketing. *Journal of Marketing, 45,* 79–88.

Blumer, H. (1969). *Symbolic interactionism: Perspective and method.* Englewood Cliffs, NJ: Prentice-Hall.

Bočkarjova, M., van der Veen, A., & Geurts, P. A. T. M. (2009). A PMT-TTM model of protective motivation for flood danger in the Netherlands. ITC Working Papers Series–Paper 3. Enschede, Netherlands: International Institute for Geo-Information Science and Earth Observation.

Bostrom, A., Fischhoff, B., & Granger Morgan, M. (1992). Characterizing mental models of hazardous processes: A methodology and an application to radon. *Journal of Social Issues, 48*(4), 85–100.

Breakwell, G. M. (2000). Risk communication: Factors affecting impact. *British Medical Bulletin, 56*(1), 110–120.

Brenkert-Smith, H., Champ, P. A., & Flores, N. (2006). Insights into wildfire mitigation decisions among wildland-urban interface residents. *Society and Natural Resources, 19*(8), 759–768.

Bright, A. D., & Manfredo, M. J. (1995). The quality of attitudinal information regarding natural resource issues: The role of attitude-strength, importance, and information. *Society & Natural Resources, 8*(5), 399–414.

Brislin, R. (1986). The wording and translation of research instruments. In W. Lonner & J. Berry (Eds.), *Field methods in cross-cultural research* (pp. 137–164). Newbury Park, NJ: Sage.

Brislin, R. (2000). *Understanding culture's influence on behavior.* Melbourne: Wadsworth.

Brun, W. (1992). Cognitive components in risk perception: Natural versus man-made risks. *Journal of Behavioral Decision Making, 5,* 117–132.

Budescu, D. V., Broomell, S., & Por, H. H. (2009). Improving communications of uncertainly in the reports of the intergovernmental panel on climate change. *Psychological Science, 20,* 299–308.

Burby, R. J., Deyle, R. E., Godschalk, D. R., & Olshansky, R. B. (2000). Creating hazard resilient communities through land-use planning. *Natural Hazards Review, 1*(2), 99–106.

Burger, J. M., & Palmer, M. L. (1992). Changes in generalization of unrealistic optimism following experiences with stressful events: Reactions to the 1989 California earthquake. *Personality and Social Psychology Bulletin, 18*(1), 39–43.

Burr, V. (1995). *An introduction to social constructionism.* London: Routledge.

Burton, I., Kates, R. W., & White, G. F. (1993). *The environment as hazard* (2nd ed.). New York: Guildford Press.

Business Continuity Institute. (2002). Good practice in business continuity management. *Continuity, 6,* 2.

Caballero, A., Carrera, P., Sanchez, F., Munoz, D., & Blanco, A. (2003). La experiencia emocional como predictor do los comportamientos de reisgo. *Psicothema, 15*(3), 427–432.

Cairns Post. (2012). Tsunami risk on the Reef. Retrieved December 22, 2012, from http://www.cairns.com.au/article/2012/12/21/237797_local-news.html

Carroll, M. S., Cohn, P. J., Seesholtz, D. N., & Higgins, L. L. (2005). Fire as a galvanising and fragmenting influence on communities: The case of the Rodeo-Chediski Fire. *Society and Natural Resources, 18,* 301–320.

Carter-Pokras, O., Zambrana, R. E., Mora, S. E., & Aaby, K. A. (2007). Emergency preparedness: Knowledge and perceptions of Latin American immigrants. *Journal of Health Care for the Poor and Underserved, 18*(2), 465–481.

Carvalho, A. (2007). Ideological cultures and media discourses on scientific knowledge: Rereading news on climate change. *Public Understanding of Science, 16,* 223–243.

Chaiken, S. (1980). Heuristic versus systematic information processing and the use of source versus message cues in persuasion. *Journal of Personality and Social Psychology, 39,* 752–766.

Chaiken, S., & Trope, Y. (1999). *Dual-process theories in social psychology.* New York: Guilford.

Charleson, A.W., Cook, B., & Bowering, G. (2003). *Assessing and increasing the level of earthquake preparedness in Wellington homes.* Proceedings of the 7th Pacific Conference on Earthquake Engineering. Wellington, New Zealand Society for Earthquake Engineering.

Chess, C., Salomone, K. L., & Hance, B. J. (1995). Improving risk communication in government: Research priorities. *Risk Analysis, 15*(2), 127–135.

Childs, I. (2008). Emergence of new volunteerism: Increasing community resilience to natural disasters in Japan. In K. Gow & D. Paton (Eds.), *The phoenix of natural disasters: Community resilience.* New York: Nova.

Clarke, L., & Short, J. F. J. (1993). Social organisation and risk: Some current controversies. *Annual Review of Sociology, 19,* 375–399.

Cochran, E. S., Vidale, J. E., & Tanak, S. (2004). Earth tides can trigger shallow thrust fault earthquakes. *Science, 306*(5699), 1164–1166.

Coley, R. L., Kuo, F. E., & Sullivan, W. C. (1997). Where does community grow? The social context created by nature in urban public housing. *Environment & Behavior, 29,* 468–492.

Comfort, L. K. (1994). Risk and resilience: Inter-organizational learning following the Northridge earthquake of 17 January 1994. *Journal of Contingencies and Crisis Management, 2*(3), 157–170.

Committee on Disaster Research in the Social Sciences (CDRSS). (2006). Facing hazards and disasters: Understanding human dimensions. In N. R. C. Committee on Disaster Research in the Social Sciences: Future Challenges and Opportunities (Eds.) Retrieved January 14, 2010, from http://www.nap.edu/catalog/11671.html

Cottrell, A., Bushnell, S., Spillman, M., Newton, J., Lowe, D., & Balcombe, L. (2008). Community perceptions of bushfire risk. In J. Handmer & K. Haynes (Eds.), *Community Bushfire Safety* (pp. 205). Melbourne: CSIRO Publishing.

Coutu, D. L. (2002). How resilience works. *Harvard Business Review, May,* 46–55.

Cowan, J., McClure, J., & Wilson, M. (2002). What a difference a year makes: How immediate and anniversary media reports influence judgments about earthquake. *Asian Journal of Social Psychology, 5,* 169–185.

Crozier, M., McClure, J., Vercoe, J., & Wilson, M. (2006). The effects of land zoning information on judgments about earthquake damage. *Area, 38*(2), 143–152.

Danesi, M. (2002). *Understanding media semiotics.* London: Arnold.

Dahlhamer, J. M., & D'Souza, M. J. (1997). Determinants of business-disaster preparedness in two U.S. metropolitan areas. *International Journal of Mass Emergencies and Disasters, 15,* 265–281.

Dahlhamer, J. M., & Tierney, K. J. (1998). Rebounding from disruptive events: Business recovery following the Northridge Earthquake. *Sociological Spectrum, 18,* 121–141.

Dake, K. (1992). Myths of nature: Culture and the social construction of risk. *Journal of Social Issues, 48*(4), 21–37.

Dalton, J. H., Elias, M. J., & Wandersman, A. (2001). *Community psychology.* Belmont, CA: Wadsworth.

Dalton, J. H., Elias, M. J., & Wandersman, A. (2005). *Community psychology* (2nd ed.). Belmont, CA: Wadsworth.

Dalton, J. H., Elias, M. J., & Wandersman, A. (2007). *Community psychology: Linking individuals and communities* (2nd ed.). Belmont, CA.: Wadsworth.

Damasio, A. (1994). *Descartes' error: Emotion, reason, and the human brain.* New York: Avon.

Daniel, T. C. (2007a). Managing individual response: Lessons from public health risk behavioral research. In W. E. Martin, C. Raish, & B. Kent (Eds.), *Wildfire risk: Human perceptions and management implications* (pp. 103–116). Washington, DC: Resources for the Future.

Daniel, T. C. (2007b), Perceptions of wildfire risk. In T. C. Daniel, M. S., Carroll, & C. Moseley (Eds.), *People, fire, and forests: A synthesis of wildfire social science* (pp. 9–36). Oregon State University Press.

Davies, H., & Walters, M. (1998). Do all crises have to become disasters? Risk and risk mitigation. *Disaster Prevention and Management, 7,* 396–400.

Davis, M. S., Ricci, T., & Mitchell, L. M. (2005). Perceptions of risk for volcanic hazards at Vesuvio and Etna, Italy. *The Australasian Journal of Disaster and Trauma Studies, 1.* http://www.massey.ac.nz/~trauma/issues/2005-1/davis.htm

Deaux, K., & Philogene, G. (Eds). (2001). *Representations of the social.* Oxford: Blackwell.

DeMan, A., & Simpson-Housley, P. (1988). Correlates of response to two potential hazards. *The Journal of Social Psychology, 128,* 385–391.

de Kwaadsteniet, E. W. (2007). *Uncertainty in social dilemmas.* Unpublished doctoral dissertation, Leiden University, The Netherlands.

Denzin, N. K. (1992). *Symbolic interactionism and cultural studies: The politics of interpretation.* Oxford: Blackwell.

Diekmann, A., & Preisendörfer, P. (1992). Personliches Umweltverhalten: Diskrepanzen Zwischen Anspruch und Wirklichkeit. *Kölner Zeitschrift Für Soziologie und Sozialpsychologie, 44,* 226–251.

Diener, E., & Suh, E. M. (2000). *Culture and subjective well-being.* Cambridge, MA: MIT Press.

Dietz, T., Gardner, G. T., Gilligan, J., Stern, P. C., & Vandenbergh, M. P. (2009). Household actions can provide a behavioral wedge to rapidly reduce carbon emissions. *PNAS (Proceedings of the National Academy of Sciences), 106,* 18452–18456.

DiPasquale, D., & Glaeser, E. L. (1999). Incentives and social capital: Are homeowners better citizens? *Journal of Urban Economics, 45*(2), 354–384.

Doepal, D. (1991). Crisis management: The psychological dimension. *Industrial Crisis Quarterly, 5,* 177 – 188.

Doll, J., & Ajzen, I. (1992). Accessibility and stability of predictors in the theory of planned behaviour. *Journal of Personality and Social Psychology, 63,* 754–765.

Dominey-Howes, D., & Minos-Minopoulos, D. (2004). Perceptions of hazard and risk on Santorini. *Journal of Volcanology and Geothermal Research, 137*(4), 285–310.

Donovan, J. L., & Blake, D. R. (1992). Patient non-compliance: Deviance or reasoned decision-making? *Social Science and Medicine, 34*(5), 507–513.

Dooley, D., Catalano, R., Mishra, S., & Serxner, S. (1992). Earthquake preparedness: Predictors in a community Survey. *Journal of Applied Social Psychology, 22,* 451–470.

Dow, K., & Cutter, S. L. (2000). Public orders and personal opinions: household strategies for hurricane risk assessment. *Environmental Hazards, 2,* 143–155.

Duitch, D., & Oppelt, T. (1997). Disaster and contingency planning: A practical approach. *Law Practice Management, 23,* 36–39.

Duncan, T. E., Duncan, S. C., Okut, H., Strycker, L. A., & Hix-Small, H. (2003). A multilevel contextual model of neighbourhood collective efficacy. *American Journal of Community Psychology, 32,* 245–252.

Duval, T. S., & Mulilis, J.-P. (1995). A person-relative-to-event (PrE) approach to negative threat appeals and earthquake preparedness: A field study. *Journal of Applied Social Psychology, 29*(3), 495–516.

Earle, T. C. (2004). Thinking aloud about trust: A protocol analysis of trust in risk management. *Risk Analysis, 24,* 169–183.

Earle, T. C., & Cvetkovich, G. T. (1995). *Social trust: Towards a cosmopolitan society.* Westport, CT: Praeger.

Eilam, O., & Suleiman, R. (2004). Cooperative, pure, and selfish trusting: Their distinctive effects on the reaction of trust recipients. *European Journal of Social Psychology, 34,* 729–738.

Elias, N., & Scotson, J. L. (1974). Cohesion, conflict and community character. In C. Bell, H. Newby, & N. Elias (Eds.), *The sociology of community: A selection of readings* (Illustrated ed., pp. 14). New York: Routledge.

Elliot, D., Swartz, E., & Herbane, B. (2002). *Business continuity management.* New York: Routledge.

Emdad Haque, C. (2000). Risk assessment, emergency preparedness and response to hazards: The case of the 1997 Red River Valley Flood, Canada. *Natural Hazards, 21*(2–3), 225–245.

Eng, E., & Parker, E. (1994). Measuring community competence in the Mississippi Delta: The interface between program evaluation and empowerment. *Health Education Quarterly, 21*(2), 199–220.

Environment Canterbury (ECan) (2004). *Solid Facts on Christchurch Liquefaction.* Retrieved December 1, 2012, from http://ecan.govt.nz/publications/General /solid-factschristchurch-liquefaction.pdf

Epstein, S. (1994). Integration of the cognitive and the psychodynamic unconscious. *American Psychologist, 49,* 709–724.

Etkin, D. (1999). Risk transference and related trends: Driving forces towards more mega-disasters. *Environmental Hazards, 1,* 69–75.

Evans, W. D. (2006). How social marketing works in healthcare. *BMJ, 332,* 1207–1210.

Faulkner, H. & Ball, D. (2007). Environmental hazards and risk communication. *Environmental Hazards, 7,* 71–78.

Finnis, K. (2004). *Creating a resilient New Zealand.* Wellington: Ministry of Civil Defence and Emergency Management.

Finucane, M. L. (2002). Mad cows, mad corn and mad communities: The role of socio-cultural factors in the perceived risk of genetically-modified food. *Proceedings of the Nutrition Society, 61*(1), 31–37.

Finucane, M. L., Alhakami, A., Slovic, P., & Johnson, S. M. (2000). The affect heuristic in judgments of risks and benefits. *Journal of Behavioral Decision Making, 13*(1), 1–17.

Fischhoff, B. (1995). Risk perception and communication unplugged: Twenty years of process. *Risk Analysis, 15*(2), 137–145.

Fischhoff, B., Slovic, P., & Lichtenstein, S. (1982). Lay foibles and expert fables in judgements about risk. *The American Statistician, 36*(3), 240–255.

Fisher, R. V., Heiken, G., & Hulen, J. B. (1998). *Volcanoes: Crucibles of change.* Princeton, NJ: Princeton University Press

Flick, U. (1998). Everyday knowledge in social psychology. In U. Flick (Ed.), *The psychology of the social* (pp. 41–59). Cambridge, UK: Cambridge University Press.

Flin, R. (1996). *Sitting in the hot seat: Leaders and teams for critical incident management.* Chichester: John Wiley & Sons.

Flora, J. L. (1998). Social capital and communities of place. *Rural Sociology, 63*(4), 481–506.

Floyd, D..L., Prentice-Dunn, S., & Rogers, R. W. (2000). A meta-analysis of research on protection motivation theory. *Journal of Applied Social Psychology, 30,* 407–429.

Flynn, J., Slovic, P., & Kunreuther, H. (Eds.). (2001). *Risk, media and stigma: Understanding public challenges to modern science and technology.* London: Earthscan

Flynn, J., Slovic, P., Mertz, C. K., & Carlisle, C. (1999). Public support for earthquake risk mitigation in Portland, Oregon. *Risk Analysis, 19,* 205–216.

Foddy, M., & Dawes, R. (2008). Group-based trust in social dilemmas. In A. Biel, D. Eek, T. Garling, & M. Gustafsson (Eds.), *New issues and paradigms in research on social dilemmas* (pp. 57–71). New York: Springer.

Folke, C., Colding, J., & Berkes, F. (2003). Synthesis: Building resilience and adaptive capacity in social-ecological systems. In F. Berkes., J. Colding, & C. Folke (Eds.), *Navigating social-ecological systems: Building resilience for complexity an change.* Cambridge: Cambridge University Press.

Forrest, R., & Kearns, A. (2001). Social cohesion, social capital and the neighbourhood. *Urban Studies, 38*(12), 2125–2143.

Fox, C. R., & Irwin, J. R. (1998). The role of context in the communication of uncertain beliefs. *Basic and Applied Social Psychology, 20*(1), 57–70.

Frandsen, M., Paton, D., & Sakariassen, K. (2011). Fostering community bushfire preparedness through engagement and empowerment. *Australian Journal of Emergency Management, 26,* 23–30.

Frandsen, M., Paton, D., Sakariassen, K., & Killalea, D. (2012). Nurturing community wildfire preparedness from the ground up: Evaluating a community engagement initiative. In D. Paton & F. Tedim (Eds.), *Wildfire and community: Facilitating preparedness and resilience* (pp. 260–280). Springfield, IL: Charles C Thomas.

Frederick, S., Loewenstein, G., & O'Donoghue, T. (2002). Time discounting and time preference: A critical review. *Journal of Economic Literature, 40,* 351–401.

Gaddy, G. D., & Tanjong, E. (1987?). Earthquake coverage by the Western press. *Journal of Communication, 36,* 105–112.

Gergen, K. J. (1985). The social constructionist movement in modern psychology. *American Psychologist, 40,* 266–275.

Gerstenberger, M., Cubrinovski, M., McVerry, G., Stirling, M., Rhoades, D., Bradley, B., Langridge, R., Webb, T., Peng, B., Pettinga, J., Berryman, K., & Brackley, H. (2011). *Probabilistic assessment of liquefaction potential for Christchurch in the Next 50 Years* (GNS Science Report 2011/15). Wellington, New Zealand: Institute of Geological and Nuclear Sciences.

Gifford, R. (1976). Environmental numbness in the classroom. *Journal of Experimental Education, 44*(3), 4–7.

Gifford, R., Iglesias, F., & Casler, J. (2009, June). *Psychological barriers to pro-environmental behavior: The development of a scale.*

Gifford, R., Scannell, L., Kormos, C., Smolova, L., Biel, A., Boncu, S., et al. (2009). Temporal pessimism and spatial optimism in environmental assessments: An 18-nation study. *Journal of Environmental Psychology, 29,* 1–12.

Godschalk, D. R., Beatley, T., Berke, P., Brower, D. J., & Edward, J. K. (1999). *Natural hazard mitigation: Recasting disaster policy and planning.* Washington, DC: Island Press.

Goodman, H., & Cottrell, A. (2012). Responding to a fire threat: Gender roles: dependency and responsibility. In D. Paton & F. Tedim (Eds.). *Wildfire and community: Facilitating preparedness and resilience* (pp. 281–299) Springfield, IL: Charles C Thomas.

Greening L., & Dollinger, S. J. (1992). Illusions (and shattered illusions) of invulnerability: Adolescents in natural disaster. *Journal of Traumatic Stress, 5,* 63–75.

Groopman, J. (2004). *The anatomy of hope.* New York: Random House.

Gordon, R. (2004). The social system as site of disaster impact and resource for recovery. *Australian Journal of Emergency Management, 19*(4), 16–22.

Graffy, E. A., & Booth, N. L. (2008). Linking environmental risk assessment and communication: An experiment in co-evolving scientific and social knowledge. *International Journal of Global Environmental Issues, 8*(1–2), 132–146.

Grant, N. K. (1996). Emergency management training and education for public administration. In R. T. Styles & W. L. Waugh (Eds.), *Disaster management in the US and Canada: The politics, policymaking, administration and analysis of emergency management* (2nd ed.) (pp. 313–325). Springfield, IL: Charles C Thomas.

Gregg, C. E., Houghton, B. F., Paton, D., Swanson, D. A., & Johnston, D. M. (200). Community preparedness for lava flows from Mauna Loa and Hualalai volcanoes, Kona, Hawaii. *Bulletin of Volcanology, 66,* 531–540.

Gregg, C. E., Houghton, B. F., Paton, D., Johnston, D. M., & Yanagi, B. (2007). Tsunami warnings: Understanding in Hawaii. *Natural Hazards, 40,* 71–87.

Gregg, C. E., Houghton, B. F., Paton, D., Swanson, D. A., Lachman, R., & Bonk, W. J. (2008). Hawaiian cultural influences on support for lava flow hazard mitigation measures during the January 1960 eruption of Kilauea volcano, Kapoho, Hawaii. *Journal of Volcanology and Geothermal Research, 172,* 300–307.

Gregg, C. E., & Houghton, B. F. (2006). Natural Hazards. In D. Paton & D. Johnston (Eds.), *Disaster resilience: An integrated approach* (pp.19–39). Springfield, IL: Charles C Thomas.

Grier, S., & Bryant, C. A., (2005). Social marketing in public health. *Annual Review of Public Health, 26,* 319–339.

Grothmann, T., & Reusswig, F. (2006). People at risk of flooding: Why some residents take precautionary action while others do not. *Natural Hazards, 38*(1–2), 101–120.

Guion, D. T., Scammon, D. L., & Borders, A. L. (2007). Weathering the storm: A social marketing perspective on disaster preparedness and response with lessons from Hurricane Katrina. *Journal of Public Policy & Marketing, 26,* 20–32.

Gunderson, L. H., Holling, C. S., & Light, S. S. (1995). *Barriers and bridges to the renewal of ecosystems and organizations.* New York: Columbia University Press.

Halpern-Felsher, B. L., Millstein, S. G., Ellen, J. M., Adler, N. E., Tschann, J. M., & Biehl, M. (2001). The role of behavioral experience in judging risks. *Health Psychology, 20*(2), 120–126.

Hannigan, A. J. (2006). *Environmental sciology: A social constructionist perspective* (2nd ed.). London: Routledge.

Hansson, S. O. (2007). Social decisions about risk and risk-taking. *Social Choice and Welfare, 29*(4), 649–663.

Hardeman, W., Johnston, M., Johnston, D. W., Bonnet, D., Wareham, N. J., & Kinmonth, A. L. (2002). Application of the theory of planned behaviour in behaviour change interventions: A systematic review. *Psychology and Health, 17*(2), 123–158.

Hardin, C. D., & Higgins, E. T. (1996). Shared reality: How social verification makes the subjective objective. In R. M. Sorrentino & E. T. Higgins (Eds.), *Motivation and cognition* (Vol. 3). New York: Guildford Press.

Hardin, G. (1968). Tragedy of commons. *Science, 162,* 1243–1247.

Harrison, M. I., & Shirom, A. (1999). *Organizational diagnosis and assessment.* Thousand Oaks, CA: Sage.

Hassol, S. J. (2008). Improving how scientists communicate about climate change. *Eos, 89,* 106–107.

Hastings, G., & McDermott, L. (2006). Putting social marketing into practice. *BMJ, 332,* 1210–1211.

Hastings, G., Stead, M., & Webb, J. (2002). Fear appeals in social marketing: Strategic and ethical reasons for concern. *Psychology & Marketing, 21,* 961–986

Heller, K., Alexander, D. B., Gatz, M., Knight, B. G., & Rose, T. (2005). Social and personal factors as predictors of earthquake preparation: The role of support provision, negative affect, age, and education. *Journal of Applied Social Psychology, 35,* 399–422.

Helweg-Larsen, M. (1999). (The Lack of) optimistic bias in response to the 1994 Northridge earthquake: The role of personal experience. *Basic and Applied Social Psychology, 21,* 119–129.

Higgins, E. T. (2000). Making a good decision: Value from fit. *American Psychologist, 55,* 1217–1230.

Hill, S. D., & Thompson, D. (2006). Understanding managers' views of global environmental risk. *Environmental Management, 37*(6), 773–787.

Hilton, D. J., Mathes, R. H., & Trabasso, T. R. (1992). The study of causal explanation in natural language: Analysing reports of the Challenger disaster in *The New York Times.* In M. L. McLaughlin, M. J. Cody, & S. J. Read (Eds.), *Explaining one's self to others* (pp. 41–59). Hillsdale, NJ: Lawrence Erlbaum.

Hine, D. W., & Gifford, R. (1996). Individual restraint and group efficiency in commons dilemmas: The effects of two types of environmental uncertainty. *Journal of Applied Social Psychology, 26,* 993–1009.

Hiroi, O., Mikami, S., & Miyata, K. (1985). A study of mass media reporting in emergencies. *International Journal of Mass Emergencies, 3,* 21–49.

Hodgson, R. W. (2007). Emotions and sense making in disturbance: Community adaptation to dangerous environments. *Human Ecology Review, 14*(2), 233–242.

Hofstede, G. (2001). *Culture's consequences: Comparing values, behaviors, institutions and organizations across nations.* Thousand Oaks, CA: Sage.

Holstein, J. A., & Miller, G. (Eds.). (2006). *Reconsidering social constructionism: Debates in social problem theory.* New York: Aldine Transaction.

Hsee, C. K., & Kunreuther, H. (2000). The affection effect in insurance decisions. *Journal of Risk and Uncertainty, 20,* 141–159.

Hummon, D. M. (1992). Community attachment: Local sentiment and sense of place. *Human Behavior & Environment: Advances in Theory & Research, 12,* 253–278.

Hurnen, F., & McClure, J. (1997). The effect of increased earthquake knowledge on perceived preventability of earthquake damage. *Australasian Journal of Disaster and Trauma Studies, 3.* http://www.massey.ac.nz/~trauma/issues/1997-3/mcclure1.htm

Indian, J. (2008). The concept of local knowledge in rural Australian fire management. In J. Handmer & K. Haynes (Eds.), *Community bushfire safety* (p. 205). Melbourne: CSIRO Publishing.

International Herald Tribune. (1995, February 2). p. 2.

Jackson, E. L. (1981). Response to earthquake hazard: The West Coast of North America. *Environment and Behavior, 13,* 387–416.

Jacobson, S. K., Monroe, M. C., & Marynowski, S. (2001). Fire at the wildland interface: The influence of experience and mass media on public knowledge, attitudes and behavioural intentions. *Wildlife Society Bulletin, 29*(3), 929–937.

Jakes, P. J. (2002). *Homeowners, communities and wildfire: Science findings from the National Fire Plan.* Paper presented at the Choices and Consequences: Natural Resources and Societal Decision-Making Conference, Bloomington, Indiana, USA.

Jakes, P. J., Nelson, K., Lang, E., Monroe, M., Agrawal, S., Kruger, L., & Strutevant, V. (2003). A model for improving community preparedness for wildfire. In P. J. Jakes (Ed.), *Homeowners, communities, and wildfire: Science findings from the national fire plan* (pp. 4–9). Gen. Tech.Rep. NC-231–Proceedings of the Ninth International Symposium on Social and Resource Management 2002, Bloomington, IN: St. Paul, MN: U.S. Department of Agriculture, Forest Service, North Central Research Station.

Jang L., & LaMendola, W. (2006). The Hakka Spirit as a predictor of resilience. In D. Paton & D. Johnston (Eds.), *Disaster resilience: An integrated approach* (pp. 174–189). Springfield, IL: Charles C Thomas.

Jardine, C. G. (2008a). Considerations in planning for successful risk communication. In B. Everitt & E. Melnick (Eds.), *Encyclopedia of quantitative risk analysis and assessment* (pp. 5–10). London: John Wiley & Sons.

Jardine, C. G. (2008b). The role of risk communication in a comprehensive risk-management approach. In B. Everitt & E. Melnick (Eds.), *Encyclopedia of quantitative risk analysis and assessment* (pp. 71–86). London: John Wiley & Sons.

Jardine, C. G. (2008c). Stakeholder participation in risk-management decision making. In B. Everitt & E. Melnick (Eds.), *Encyclopedia of quantitative risk analysis and assessment* (pp. 1–4). London: John Wiley & Sons.

Johnson, L. R., Johnson-Pynn, J. S., & Pynn, T. (2007). Youth civic engagement in China: Results from a program promoting environmental activism. *Journal of Adolescent Research, 22,* 355–386.

Johnson-Laird, P. N. (1983). *Mental models: Towards a cognitive science of language, inference and consciousness* (5th ed.). Cambridge, MA: Harvard University Press.

Johnston, D., Bebbington, M. S., Lai, C.-D., Houghton, B. F., & Paton, D. (1999). Volcanic hazard perceptions: comparative shifts in knowledge and risk. *Disaster Prevention and Management, 8*(2), 118–126.

Johnston, D., Paton, D., Crawford, G. L., Ronan, K., Houghton, B., & Buergelt, P. (2005). Measuring tsunami preparedness in Coastal Washington, United States. *Natural Hazards, 35*(1), 173–184.

Jones, B. D. (1999). Bounded rationality. *Annual Review of Political Science, 2,* 297–321.

Kahneman, D. (2003). A Perspective on judgment and choice: Mapping bounded rationality. *American Psychologist, 58*(9), 697–720.

Kee, H., & Knox, R. T. (1970). Conceptual and methodological considerations in the study of trust and suspicion. *Journal of Conflict Resolution, 14,* 357–365.

Keller, C., Siegrist, M., & Gutscher, H. (2006). The role of the affect and availability heuristics in risk communication. *Risk Analysis, 26*(3), 631–639.

Kelley, H. H. (1967). Attribution in social psychology. *Nebraska Symposium on Motivation, 15,* 192–238.

King, D. (2001). Uses and limitations of socioeconomic indicators of community vulnerability to natural hazards: Data and disasters in Northern Australia. *Natural Hazards, 24*(2), 147–156.

King, D. (2008). Reducing hazard vulnerability through local government engagement and action. *Natural Hazards, 47*(3), 497–508.

Kohler, H. P., Behrman, J. R., & Watkins, S. C. (2007). Social networks and HIV/AIDS risk perceptions. *Demography, 44*(1), 1–33.

Koriat, A., Lichtenstein, S., & Fischhoff, B. (1980). Reasons for confidence. *Journal of Experimental Psychology: Human Learning and Memory, 6,* 107–118.

Kraft, P., Rise, J., Sutton, S., & Røysamb, E. (2005). Perceived difficulty in the theory of planned behaviour: Perceived behavioural control or affective attitude? *British Journal of Social Psychology, 44*(3), 479–496.

Kunruether, H., & Pauly, M. (2004). Neglecting disaster: Why don't people insure against large losses. *The Journal of Risk and Uncertainty, 28*(1), 5–21.

Kumagai, Y., Bliss, J. C., Daniels, S. E., & Carroll, M. S. (2004). Real time research on causal attribution of wildfire: An exploratory multiple-methods approach. *Society and Natural Resources, 17*(2), 113–127.

Kasperson, R. E., Renn, O., Slovic, P., Brown, S., Emel, J., Goble, R., et al. (1988). The social amplification of risk: A conceptual framework. *Risk Analysis, 8,* 177–187.

Kok, M. T. J., & de Coninck, H. C. (2007). Widening the scope of policies to address climate change: Directions for mainstreaming. *Environmental Science and Policy, 10,* 587–599.

Kuo, F. E., Sullivan, W. C., Coley, R. L., & Brunson, L. (1998). Fertile ground for community: Inner-city neighborhood common spaces. *American Journal of Community Psychology, 26,* 823–851.

Kuo, F. E., & Sullivan, W. C. (2001). Aggression and violence in the inner city: Impacts of environment via mental fatigue. *Environment & Behavior, 33*(4), 543–571.

Lachman, R., Tatsuoka, M., & Bonk, W. J. (1961). Human behavior during the tsunami of May 23, 1960. *Science, 133*(3462), 1405–1409.

Lamontaigne, M., & La Rochelle, S. (2000). Earth scientists can help people who fear earthquakes. *Seismological Research Letters, 70,* 1–4.

Lasker, R. D. (2004). *Redefining readiness: Terrorism planning through the eyes of the public.* New York: The New York Academy of Medicine.

Latimer, A. E., & Martin-Ginis, K. A. (2005). The importance of subjective norms for people who care what others think of them. *Psychology and Health, 20,* 53–62.

Lehman, D., & Taylor, S. E. (1987). Date with an earthquake: Coping with a probable, unpredictable disaster. *Personality and Social Psychology Bulletin, 13,* 546–555.

Leiserowitz, A. (2007). *American opinions on global warming: A Yale University / Gallup/ ClearVision Institute Poll.* New Haven, CT: Yale School of Forestry and Environmental Studies.

Levene, L. (2004). *Taming the beast–Managing business risk.* London: Lloyd's of London.

Ley, D., & Murphy, P. (2001). Immigration in gateway cities: Sydney and Vancouver in comparative perspective. *Progress in Planning, 55*(3), 119–194.

Lichterman, J. D. (2000). A 'community as resource' strategy for disaster response. *Public Health Reports, 115*(2–3), 262–265.

Lichtenstein, S., & Fischhoff, B. (1977). Do those who know more also know more about how much they know? *Organizational Behavior and Decision Processes, 20,* 159–183.

Lindell, M. K. (1994). Perceived characteristics of environmental hazards. *International Journal of Mass Emergencies and Disasters, 12,* 303–326.

Lindell, M. K., & Hwang, S. N. (2008). Households' perceived personal risk and responses in a multi-hazard environment. *Risk Analysis, 28,* 539–556.

Lindell, M. K., & Perry, R. (1992). *Behavioural foundations of community emergency planning.* Washington: Taylor & Francis.

Lindell, M. K., & Perry, R. W. (2000). Household adjustment to earthquake hazard: A review of research. *Environment and Behaviour, 32*(4), 461–501.

Lindell, M. K., & Perry, R. W. (2004). *Communicating environmental risk in multiethnic communities.* Thousand Oaks, CA: Sage.

Lindell, M. K., & Whitney, D. J. (2000). Correlates of household seismic hazard adjustment adoption. *Risk Analysis, 20*(1), 13–25.

Lindell, M. K., Arlikatti, S., & Prater, C. S. (2009). Why do people do what they do to protect against earthquake risk: Perception of hazard adjustment attributes. *Risk Analysis, 29,* 1072–1088.

Lindell, M. K., Prater, C. S., & Perry, R. W. (2006). *Fundamentals of emergency management.* Emmitsburg, MD: Federal Emergency Management Agency.

Linville, P. W., & Fischer, G. W. (1991). Preferences for separating and combining events: A social application of prospect theory and the mental accounting model. *Journal of Personality and Social Psychology, 60,* 5–33.

Lion, R., Meertens, R. M., & Bot, I. (2002). Priorities in information desire about unknown risks. *Risk Analysis, 22,* 765–776.

Lippke, S., & Ziegelmann, J. P. (2008). Theory-based health behavior change: Developing, testing and applying theories for evidence-based interventions. *Applied Psychology, 57*(4), 698–716.

Lippke, S., Wiedermann, A. U., Ziegelmann, J. P., Reuter, T., & Schwarzer, R. (2009). Self-efficacy moderates the mediation of intentions into behavior via plans. *American Journal of Health Behavior, 33*(5), 521–529.

Lippke, S., & Ziegelmann, J. P. (2008). Theory-based health behavior change: Developing, testing and applying theories for evidence-based interventions. *Applied Psychology, 57*(4), 698–716.

Lister, K. (1996). Disaster continuity planning. *Chartered Accountants Journal of New Zealand, 75,* 72–73.

Loewenstein, G. F., Hsee, C. K., Weber, E. U., & Welch, N. (2001). Risk as feelings. *Psychological Bulletin, 127*(2), 267–286.

Lopes, R. (1992). *Public perception of disaster preparedness [resentations using disaster images.* Washington DC: The American Red Cross.

Lorenzoni, I., Nicholson-Cole, S., & Whitmarsh, L. (2007). Barriers perceived to engaging with climate change among the UK public and their policy implications. *Global Environmental Change, 17,* 223–233.

Low, S. M., & Altman, I. (1992). Place attachment: A conceptual inquiry. In I. Altman, & S. M. Low (Eds.). *Place attachment.* New York: Plenum Press.

Luhmann, N. (1979). *Trust and power.* Chichester: Wiley.

Lupton, D. (1999). *Risk.* London: Routledge.

Lupton, D., & Tulloch, J. (2002). Risk is part of your life: Risk epistemologies among a group of Australians. *Sociology, 36*(2), 317–334.

Macy, J., & Brown, M. Y. (1998). *Coming back to life: Practices to reconnect our lives, our world.* Gabriola Island, British Columbia: New Society.

MacGregor, D., Slovic, P., Mason, R. G., Detwiler, J., Binney, S. J., & Dodd, B. (1994). Perceived risks of radioactive waste transport through oregon: Results of a statewide survey. *Risk Analysis,* 14, 5–14.

Maddux, J. E., . & Rogers, R. W. (1983). Protection motivation and self-efficacy: A revised theory of fear appeals and attitude change. *Journal of Experimental Social Psychology, 19,* 469–479.

Maibach, E., & Holtgrave, D. R. (1995). Advances in public health communication. *Annual Review of Public Health, 16,* 219–238.

Mamula-Seadon, L., Selway, K, & Paton, D. (2012). Exploring resilience: Learning from Christchurch communities. *Tephra, 23,* 5–7.

Marris, C., Langford, I. H., & O'Riodan, T. (1998). A quantitative test of the cultural theory of risk perception: Comparisons with the psychometric paradigm. *Risk Analysis, 18,* 635–647.

Martin, I. M., Bender, H., & Raish, C. (2007a). What motivates individuals to protect themselves from risks: The case of wildland fires. *Risk Analysis, 27*(4), 887–900.

Martin, I. M., Bender, H.W. & Raish, C. (2007b). Making the decision to mitigate risk. In W. E. Martin, C. Raish, & B. Kent (Eds.), *Wildfire risk: Human perceptions and management implications* (pp. 117–141). Washington, DC: Resources for the Future Press.

Marks, I. M., & Matthews, A. M. (1979). Brief standard self-rating for phobic patients. *Behaviour Research and Therapy, 17,* 2633–267.

Matsumoto, D., & Juang, L. (2008). *Culture and psychology.* Belmont CA: Thompson/Wadsworth.

Mayer, R. C., Davis, J. H., & Schoorman, F. D. (1995). An integrative model of organizational trust. *Academy of Management Review, 20,* 709–734.

McCaffrey, S. (2004a). Fighting fire with education: What is the best way to reach out to homeowners? *Journal of Forestry, 102*(5), 12.

McCaffrey, S. (2004b). Thinking of wildfire as a natural hazard. *Society and Natural Resources, 17*(6), 509–516.

McClure, J. L., & Sibley, C. G. (2011). Framing effects on disaster preparation: Is negative framing more effective? *Australasian Journal of Disaster and Trauma Studies, 11.* http://www.massey.ac.nz/~trauma/issues/2011-1/mcclure.htm

McClure, J., & Velluppillai, J. (under review). The effects of news media reports on earthquake judgments: Mixed messages about the Canterbury earthquake. *Australasian Journal of Disaster and Trauma Studies.*

McClure, J. L., & Williams, S. (1996). Community preparedness: Countering helplessness and optimism. In D. Paton & N. Long (Eds.), *Psychological aspects of dis-*

aster: Impact, coping, and prevention (pp. 237–254). Palmerston North: Dunmore Press.

McClure, J., Allen, M. W., & Walkey, F. (2001). Countering fatalism: Causal information in news reports affects judgements about earthquake damage. *Basic and Applied Social Psychology, 23*(2), 109–121.

McClure, J., Sibley, C. G., & Rose, C. S. (under review). Failure to prepare for low frequency hazards: Do people discount low frequency risks if the per annum cost is held constant? *Australasian Journal of Disaster and Trauma Studies.*

McClure, J., Sutton, R. M., & Sibley, C. (2007). Listening to reporters or engineers? How different messages about building design affect earthquake fatalism. *Journal of Applied Social Psychology, 37,* 1956–1973.

McClure, J., Sutton, R. M., & Wilson, M. (2007). How information about building design influences causal attributions for earthquake damage. *Asian Journal of Social Psychology, 10,* 233–242.

McClure, J., Walkey, F., & Allen, M. (1999). When earthquake damage is seen as preventable: Attributions, locus of control and attitudes to risk. *Applied Psychology, 48*(2), 239–256.

McClure, J., White, J., & Sibley, C. G. (2009). Framing effects on preparation intentions: Distinguishing actions and outcomes. *Disaster Prevention and Management, 18,* 187–199.

McClure, J., Fischer, R., Charleson, A., & Spittal, M. J. (2009). *Clarifying why people take fewer damage mitigation actions than survival actions: How important is cost?* Wellington, NZ: Earthquake Commission.

McClure, J., Wills, C., Johnston, D., & Recker, C. (2011). How the 2010 Canterbury (Darfield) earthquake affected earthquake risk perception: Comparing citizens inside and outside the earthquake region. *Australasian Journal of Disaster and Trauma Studies, 2.* http://www.massey.ac.nz/~trauma/issues/2011-2/AJDTS _2011-2_McClure.pdf

McComas, K. A. (2003). Public meetings and risk amplification: A longitudinal study. *Risk Analysis, 23*(6), 1257–1270.

McDaniels, T. L. (1988). Chernobyl's effects on the perceived risks of nuclear power: A small sample test. *Risk Analysis, 8,* 457–461.

McGee, T. K., & Russell, S. (2003). "It's just a natural way of life . . ." an investigation of wildfire preparedness in rural Australia. *Environmental Hazards, 5*(1–2), 1–12.

McGuire, W. (2012). *Waking the giant: How a changing climate triggers earthquakes, tsunamis, and volcanoes.* London: Oxford University Press.

McIvor, D., & Paton, D. (2007). Preparing for natural hazards: Normative and attitudinal influences. *Disaster Prevention and Management, 16*(1), 79–88.

McIvor, D., Paton, D., & Johnston, D. M. (2009). Modelling community preparation for natural hazards: Understanding hazard cognitions. *Journal of Pacific Rim Psychology, 3,* 39–46.

McLennan, J., & Elliot, G. (2012). Community bushfire safety issues: Findings from interviews with residents affected by the 2009 Victorian bushfires. *Bushfire Cooperative Research Centre Fire Note, 98,* October.

McMillan, D. W., & Chavis, D. M. (1986). Sense of community: A definition and theory. *Journal of Community Psychology, 14*(1), 6–23.

McKenzie-Mohr, D. (2000). Fostering sustainable behavior through community-based social marketing. *American Psychologist, May,* 531–537.

McKenzie-Mohr, D., & Smith, W. (1999). *Fostering sustainable behavior: An introduction to community-based social marketing.* Gabriola Island, BC: New Society.

Meichenbaum, D. (1996). Stress inoculation training for coping with stressors. *The Clinical Psychologist, 49,* 4–7.

Meichenbaum, D. (2007). Stress inoculation training: A preventative and treatment approach. In P. M. Lehrer, R. L. Woolfolk, & W. S. Sime (Eds.), *Principles and practice of stress management* (3rd ed.) (pp. 3–40). New York: Guilford Press.

Miceli, R., Sotgiu, I., & Settanni, M. (2008). Disaster preparedness and perception of flood risk: A study in an alpine valley in Italy. *Journal of Environmental Psychology, 28,* 164–173.

Mileti, D. S., & Darlington, J. D. (1995). Societal response to revised earthquake probabilities in the San Francisco Bay area. *International Journal of Mass Emergencies and Disasters, 13,* 119–145.

Mileti, D. S., & Fitzpatrick, C. (1992). The casual sequence of risk communication in the Parkfield earthquake prediction experiment. *Risk Analysis. 9,* 20–28.

Mileti, D. S., & Fitzpatrick, C. (1993). *The great earthquake experiment.* Boulder, CO: Westview Press.

Mileti, D. S., & O'Brien, P. (1993). Public response to aftershock warnings. *U.S. Geological Survey Professional Paper. 1553–B,* 31–42.

Mileti, D. S., Fitzpatrick, C., & Farhar, B. C. (1992). Fostering public preparations for natural hazards: Lessons from the Parkfield earthquake prediction. *Environment, 34,* 16–20.

Mileti, D. S., Nathe, S., Gori, P., & Lemersal, E. (2004). *Public hazards communication and education: The state of the art.* Boulder, CO: Natural Hazards Research and Applications Information Center, University of Colorado at Boulder.

Milfont, T. L., & Gouveia, V. V. (2006). Time perspective and values: An exploratory study of their relation to environmental attitudes. *Journal of Environmental Psychology, 26,* 72–82.

Milinski, M., Semmann, D., Krambeck, H. J., & Marotzke, J. (2006). Stabilizing the earth's climate is not a losing game: Supporting evidence from public goods experiments. *Proceedings of the National Academy of Sciences of the United States of America, 103,* 3994–3998.

Mitroff, I. I., & Anagnos, G. (2001). *Managing crises before they happen.* New York: AMACOM.

Mulilis, J.-P., & Duval, T. S. (1995). Negative threat appeals and earthquake preparedness: A person-relative-to-event (PrE) model of coping with threat. *Journal of Applied Social Psychology, 25,* 1319–1339.

Mulilis, J.-P., Duval, T. S., & Bovalino, K. (2000). Tornado preparedness of students, non-student renters and non-student owners: Issues of PrE theory. *Journal of Applied Social Psychology, 30,* 1310–1329.

Mills, M. J. (2000). Volcanic aerosol and global atmospheric effects. In B. H. H. Sigurdsson, S. McNutt, H. Rymer, & J. Stix (Eds.), *Encyclopedia of volcanoes* (pp. 931–943). San Diego, CA: Academic Press.

Morgan, M. G., & Lave, L. (1990). Ethical considerations in risk communication practice and research. *Risk Analysis, 10*(3), 355–358.

Morrison, N. (2003). Neighbourhoods and social cohesion: Experiences from Europe. *International Planning Studies, 8*(2), 115–138.

Morrissey, S., & Reser, J. (2003). Evaluating the effectiveness of psychological preparedness advice in community cyclone preparedness materials. *Australian Journal of Emergency Management, 18*(2), 46–61.

Moscovici, S. (1984). The phenomenon of social representations. In R. M. Farr & S. Moscovici (Eds.), *Social representations* (pp. 3–69). Cambridge, UK: Cambridge University Press.

Moscovici, S. (2000). *Social representations: Explorations in social psychology.* Cambridge, UK: Polity Press.

Moser, S. C. (2007). More bad news: The risk of neglecting emotional responses to climate change information. In S. C. Moser & L. Dilling (Eds.), *Creating a climate for change* (pp. 64–80). New York: Cambridge University Press.

Moser, S. C., & Dilling, L (2004). Making climate hot: Communicating the urgency and challenge of global climate change. *Environment, 46,* 32–46.

Mulilis, J. P. (1998). Persuasive communication issues in disaster management: A review of the hazards mitigation and preparedness literature and a look towards the future. *Australian Journal of Emergency Management, 13*(1), 51–59.

Mulilis, J. P., & Duval, T. S. (1995). Negative threat appeals and earthquake preparedness: A person-relative-to-event (PrE) model of coping with threat. *Journal of Applied Social Psychology, 25*(15), 1319–1339.

Mulilis, J. P., & Lippa, R. (1990). Behavioral change in earthquake preparedness due to negative threat appeals: A test of protection motivation theory. *Journal of Applied Social Psychology, 20*(1), 619–638.

Mulilis, J. P., Duval, T. S., & Lippa, R. (1990). The effects of a large destructive local earthquake on earthquake preparedness as assessed by an earthquake preparedness scale. *Natural Hazards, 3*(4), 357–371.

National Environment Protection Council. (1999). *Guidelines on community consultation and risk communication.* Canberra: National Environment Protection Council

Nakano, L. (2005). *Community volunteers in Japan: Everyday stories of social change.* London: Routledge.

Newhall, C. G., & Punongbayan, R. S. (Eds.). (1996). *Fire and mud eruptions and lahars of Mount Pinatubo, Philippines.* Seattle, WA: University of Washington Press.

Nicholsen, S. W. (2002). *The love of nature and the end of the world.* Cambridge, MA: MIT Press.

Norenzayan, A., & Heine, S. J. (2005). Psychological universals: What are they and how can we know? *Psychological Bulletin, 131,* 763–784.

Olson, R., & Rejeski, D. (2005). *Environmentalism and the technologies of tomorrow: Shaping the next industrial revolution.* Washington, DC: Island Press.

Owen, A. J., Colbourne, J. S., Clayton, C. R. I., & Fife-Schaw, C. (1999). Risk communication of hazardous processes associated with drinking water quality–A mental models approach to customer perception, part 1–A methodology. *Water Science and Technology, 39*(10–11), 183–188.

Palm, R., Hodgson, M., Blanchard, R. D., & Lyons, D. (1990). *Earthquake insurance in California.* Boulder, CO: Westview Press.

Park, H. S. (2000). Relationships among attitudes and subjective norms: Testing the theory of reasoned action across cultures. *Communication Studies, 51*(2), 162.

Paton, D. (1989). Disasters and helpers: Psychological dynamics and implications for counselling. *Counselling Psychology Quarterly, 2,* 303–321.

Paton, D. (1997a). *Dealing with traumatic incidents in the workplace* (3rd ed). Queensland: Gull.

Paton, D. (1997b). Managing work-related psychological trauma: An organisational psychology of response and recovery. *Australian Psychologist, 32,* 46–55.

Paton, D. (2000). Emergency planning: Integrating community development, community resilience and hazard mitigation. *Journal of the American Society of Professional Emergency Planners, 7,* 109–118.

Paton, D. (2003a). Disaster preparedness: A social cognitive perspective. *Disaster Prevention and Management, 12*(3), 210–216.

Paton, D. (2003b). Stress in disaster response: A risk management approach. *Disaster Prevention and Management, 12,* 203–209.

Paton, D. (2005). *Community resilience: Integrating hazard management and community engagement.* Proceedings of the International Conference on Engaging Communities. Brisbane: Queensland Government/UNESCO.

Paton, D. (2006a). Disaster resilience: Building capacity to co-exist with natural hazards and their consequences. In D. Paton & D. Johnston (Eds.), *Disaster resilience: An integrated approach* (pp. 3–10). Springfield, IL: Charles C Thomas.

Paton, D. (2006b). Disaster resilience: Integrating individual, community, institutional and environmental perspectives. In D. Paton & D. Johnston (Eds.), *Disaster resilience: An integrated approach* (pp. 305–318). Springfield, IL: Charles C Thomas.

Paton, D. (2006c). *Measuring and monitoring resilience.* Auckland: Auckland Region CDEM Group

Paton, D. (2006d). *Promoting household community preparedness for bushfires: A review of issues that inform the development and delivery of risk communication strategies.* Melbourne: Bushfire Co-operative Research Centre.

Paton, D. (2007a). *Measuring and monitoring resilience in Auckland* (GNS Science Report 2007/18). Wellington, New Zealand: GNS Science.

Paton, D. (2007b). Preparing for natural hazards: The role of community trust. *Disaster Prevention and Management, 16*(3), 370–379.

Paton, D. (2008a). Community resilience: Integrating individual, community and societal perspectives. In K. Gow & D. Paton (Eds.), *The phoenix of natural disasters: Community resilience.* New York: Nova Science.

Paton, D. (2008b). *Modelling societal resilience to pandemic hazards in Auckland* (GNS Science Report 2008/13). Wellington, New Zealand: Institute of Geological and Nuclear Sciences.

Paton, D. (2008c). Risk communication and natural hazard mitigation: how trust influences its effectiveness. *International Journal of Global Environmental Issues, 8*(1–2), 2–16.

Paton, D. (2009). Community sustainability and natural hazard resilience: All-hazard and cross-cultural issues in disaster resilience. *The International Journal of Environmental, Cultural, Economic & Social Sustainability, 5,* 345–356.

Paton, D. (2012). *MCDEM Christchurch community resilience project report: Part 1.* Ministry of Civil Defence and Emergency Management. Wellington, New Zealand.

Paton, D., & Bishop, B. (1996a). Disasters and communities: Promoting psychosocial well-being. In D. Paton & N. Long (Eds.), *Psychological aspects of disaster: Impact, coping, and intervention* (pp. 255–268). Palmerston North, New Zealand: Dunmore Press.

Paton, D., & Buergelt, P. T. (2005, October). *Living with bushfire risk: Residents accounts of their bushfire preparedness behaviour.* Paper presented at AFAC/Bushfire CRC Conference, Auckland, New Zealand.

Paton, D., & Buergelt, P. T. (2012). Community engagement and wildfire preparedness: The influence of community diversity. In D. Paton & F. Tedim (Eds.), *Wildfire and community: Facilitating preparedness and resilience* (pp. 241–259). Springfield, IL: Charles C Thomas.

Paton, D., & Jackson, D. (2002). Developing disaster management capability: An assessment centre approach. *Disaster Prevention and Management, 11,* 115–122.

Paton, D., & Jang, L. (2011). Disaster resilience: Exploring all-hazards and cross-cultural perspectives. In D. Miller & J. Rivera (Eds.), *Community disaster recovery and resiliency: Exploring global opportunities and challenges.* London: Taylor & Francis

Paton, D., & Johnston, D. (2006). *Disaster resilience: An integrated approach.* Springfield, IL: Charles C Thomas.

Paton, D., & Johnston, D. (2008). *A means-end chain theory analysis of hazard cognitions and preparedness.* Wellington, New Zealand: Earthquake Commission.

Paton, D., & Tang, C.S. (2009). Adaptive and growth outcomes following tsunami: The experience of Thai communities following the 2004 Indian Ocean Tsunami. In Edward S. Askew & J. P. Bromley (Eds.), *Atlantic and Indian oceans: New oceanographic research* (pp. 125–140). New York: Nova Science.

Paton, D., & Tedim, F. (2012). Wildfire risk management: Building on lessons learnt. In D. Paton & F. Tedim (Eds.), *Wildfire and community: Facilitating preparedness and resilience* (pp. 323–336). Springfield, IL: Charles C Thomas.

Paton, D., & Wilson, F. (2001). Managerial perceptions of competition in knitwear producers. *Journal of Managerial Psychology, 16,* 289–300.

Paton, D., & Wright, L. (2008). Preparing for bushfires: The public education challenges facing fire agencies. In J. Handmer & K. Haynes (Eds.), *Community bushfire safety* (pp. 117–128). Canberra: CSIRO Publishing.

Paton, D., Buergelt, P. T., & Prior, T. (2008). Living with bushfire risk: Social and environmental influences on preparedness. *Australian Journal of Emergency Management, 23*(3), 41–48.

Paton, D., Johnston, D., & Houghton, B. (1998). Organisational responses to a volcanic eruption. *Disaster Prevention and Management, 7,* 5–13.

Paton, D., Johnston, D., & Johgens, R. (2003). *Napier city council hazard analysis research project. Part 2: Social vulnerability* (GNS Science Report 2003/76). Upper Hutt, New Zealand: Institute of Geological and Nuclear Sciences.

Paton, D., McClure, J., & Buergelt, P. T. (2006). Natural hazard resilience: The role of individual and household preparedness. In D. Paton & D. Johnston (Eds.), *Disaster resilience: An integrated approach* (p. 321). Springfield, IL: Charles C Thomas.

Paton, D., Millar, M., & Johnston, D. (2001). Community resilience to volcanic hazard consequences. *Natural Hazards, 24,* 157–169.

Paton, D., Smith, L., & Johnston, D. (2000). Volcanic hazards: Risk perception and preparedness. *New Zealand Journal of Psychology, 29*(2), 86–91.

Paton, D., Smith, L., & Johnston, D. (2003). *A means-end chaint theory analysis of hazard cognitions and preparedness.* Wellington, New Zealand: Earthquake Commission.

Paton, D., Smith, L., & Johnston, D. (2005). When good intentions turn bad: Promoting natural hazard preparedness. *Australian Journal of Emergency Management, 20*(1), 25–30.

Paton, D., Smith, L., & Violanti, J. (2000). Disaster response: Risk, vulnerability and resilience. *Disaster Prevention and Management, 9*(3), 173.

Paton, D., Bajek, R., Okada, N., & McIvor, D. (2010). Predicting community earthquake preparedness: A cross-cultural comparison of Japan and New Zealand. *Natural Hazards, 54,* 765–781

Paton, D., Kelly, G., Buergelt, P. T., & Doherty, M. (2006). Preparing for bushfires: Understanding intentions. *Disaster Prevention and Management, 15*(4), 566–575.

Paton, D., Smith, L., Daly, M., & Johnston, D. (2008). Risk perception and volcanic hazard mitigation: Individual and social perspectives. *Journal of Volcanology and Geothermal Research, 172*(3–4), 179–188.

Paton, D., Johnston, D. M., Bebbington, M. S., Lai, C.-D., & Houghton, B. F. (2001). Direct and vicarious experiences of volcanic hazards: Implications for risk perception and adjustment adoption. *Australian Journal of Emergency Management, Summer,* 58–63.

Paton, D., Smith, L., Johnston, D., Johnston, M., & Ronan, K. (2003). *Developing a model of intentions to predict natural hazard preparation* (EQC Research Report No. 01-479). Wellington, New Zealand: New Zealand Earthquake Commission.

Paton, D., Smith, L., Johnston, D., Johnston, M., & Ronan, K. (2004). *Developing a model to predict the adoption of natural hazard risk reduction and preparatory adjustments* (No. EQC Research Project No. 01-479). Wellington, New Zealand: Earthquake Commission.

Paton, D., Gregg, C. E., Houghton, B. F., Lachman, R., Lachman, J., Johnston, D. M., et al. (2008). The impact of the December 26th 2004 tsunami on coastal Thai communities: Assessing adaptive capacity. *Disasters, 32*(1),106–119.

Paton, D., Houghton, B. F., Gregg, C. E., Gill, D. A., Ritchie, A., McIvor, D., et al. (2008). Managing tsunami risk in coastal communities: Identifying predictors of preparedness. *Australian Journal of Emergency Management, 23*(1), 4–9.

Paton, D., Houghton, B. F., Gregg, C. E., McIvor, D., Johnston, D. M., Buergelt, P. T., et al. (2009). Managing tsunami risk: Social context influences on preparedness. *Journal of Pacific Rim Psychology, 3,* 27–37.

Perry, R.W., & Lindell, M. K. (2008). Volcanic risk perception and adjustment in a multi-hazard environment. *Journal of Volcanology and Geothermal Research, 172,* 170–178.

Pettenger, M. E. (2007). *The social construction of climate change.* London: Ashgate.

Petts, J. (2004). Barriers to participation and deliberation in risk decisions: Evidence from waste management. *Journal of Risk Research, 7*(2), 115–133.

Pfeiffer, T., & Nowak, M. A. (2006). Climate change: All in the game. *Nature, 441,* 583–584.

Pidgeon, N. F., Kasperson, R. K., & Slovic, P. (2003). *The social amplification of risk.* Cambridge, UK: Cambridge University Press.

Plous, S. (1993). *The psychology of judgment and decision making.* New York: McGraw-Hill.

Poortinga, Y. (1997). Towards convergence? In J. Berry, Y. Poortinga, & J. Pandey (Eds.), *Theory and method: Vol. 1. Handbook of cross cultural psychology* (2nd ed.) (pp. 347–387). Boston: Allyn & Bacon.

Poortinga, W., & Pidgeon, N. F. (2004). Trust, the asymmetry principle, and the role of prior beliefs. *Risk Analysis, 24,* 1475–1486.

Portes, A. (1998). Social capital: Its origins and applications in modern sociology. *Annual Review of Sociology, 24,* 1–24.

Portes, A. (1998). Social capital: Its origins and applications in modern sociology. *Annual Review of Sociology, 24,* 1–24.

Powell, M., Dunwoody, S., Griffin, R., & Neuwirth, K. (2007). Exploring lay uncertainty about an environmental health risk. *Public Understanding of Science, 16*(3), 323–342.

Prewitt Diaz, J. O., & Dayal, A. (2008). Sense ofplace: A model for community based psychosocial support programs. *Australasian Journal of Disaster and Trauma Studies, 1.* http://www.massey.ac.nz/~trauma/issues/2008-1/prewitt_diaz.htm

Prior, T., & Eriksen, C. (2012). What does being 'well prepared' for wildfire mean? In D. Paton & F. Tedim (Eds.), *Wildfire and community: Facilitating preparedness and resilience* (pp. 190–206). Springfield, IL: Charles C Thomas.

Proudley, M. (2008). Fire, families and decisions. *Australian Journal of Emergency Management, 23*(1), 37–43.

Putnam, R. D. (2000). *Bowling alone: The collapse and revival of American community.* New York: Simon and Schuster Paperbacks.

Prochaska, J. O., & DiClemente, C. C. (1982). Transtheoretical therapy: Toward a more integrative model of change. *Psychotherapy: Theory, Research and Practice, 19,* 276–288.

Prior, T. (2010). *Householder bushfire preparation: Decision making and the implications for risk communication.* Unpublished PhD dissertations, University of Tasmania, Hobart, Australia.

Recchia, V. (1999). *Risk communication and public perception of technological hazards* (Vol. 1). Milan, Italy: Fondazione Eni Enrico Mattei.

Reiss, C. L. (2004). *Risk management for small business.* Fairfax, VA: Public Entity Risk Institute.

Rich, R. C., Edelstein, M., Hallman, W. K., & Wandersman, A. H. (1995). Citizen participation and empowerment: The case of local environmental hazards. *American Journal of Community Psychology, 23,* 657–677.

Ripley, A. (2006). Floods, tornadoes, hurricanes, wildfires, earthquakes. . . . Why we don't prepare. *Time,* 49–52.

Rippl, S. (2002). Cultural theory and risk perception: A proposal for better measurement. *Journal of Risk Research, 5,* 147–165.

Robberson, M., & Rogers, R. (1988). Beyond fear appeals: Negative and positive persuasive appeals to health and self-esteem. *Journal of Applied Social Psychology, 61,* 277–287.

Rogers, E. M. (1995). *Diffusion of innovations* (4th ed.). New York: Free Press.

Rogers, R. W. (1975). A protection motivation theory of fear appeals and attitude change. *The Journal of Psychology, 91,* 93–114.

Rogers, R. W. (1983). Cognitive and physiological processes in fear appeals and attitude change: A revised theory of protection motivation. In B. L. Cacioppo & L. L. Petty (Eds.), *Social psychophysiology: A sourcebook* (pp. 153–176). London: Guilford.

Rohrmann, B. (1994). Risk perception of different societal groups: Australian findings and cross-national comparisons. *Australian Journal of Psychology, 46,* 150–163.

Rose, A., & Lim, D (2002). Business interruption losses from natural hazards: Conceptual and methodological issues in the case of the Northridge earthquake. *Environmental Hazards, 4,* 1–14.

Rosenbaum, M. (1990). *Learned resourcefulness: On coping skills, self-control, and adaptive behavior.* New York: Springer.

Rosenthal, P. H., & Sheiniuk, G. (1993). Business resumption planning: Exercising the disaster management team. *Journal of Systems Management, 44,* 12–16.

Rosenstock, I. M. (1974). Historical origins of the health belief model. *Health Education Monographs, 2,* 1–8.

Russell, L. A., Goltz, J. D., & Bourque, L. B. (1995). Preparedness and hazard mitigation actions before and after two earthquakes. *Environment & Behaviour, 27*(6), 744–770.

Sarason, S. B. (1974). *The psychological sense of community: Prospects for a community psychology.* San Francisco: Jossey-Bass.

Sattler, D. N., Preston, A., Kaiser, C. F., Olivera, V. E., Valdez, J., & Schlueter, S. (2002). Hurricane Georges: A cross-national study examining preparedness, resource loss, and psychological distress in the U. S. Virgin Islands, Puerto Rico, Dominican Republic, and the United States. *Journal of Traumatic Stress, 15,* 339–350.

Schwarzer, R. (2008). Modeling health behavior change: How to predict and modify the adoption and maintenance of health behaviors. *Applied Psychology, 57*(1), 1–29.

Searles, H. F. (1972). Unconscious processes in relation to the environmental crisis. *Psychoanalytical Review, 59,* 361–374.

Severtson, D. J., Baumann, L. C., & Brown, R. L. (2006). Applying a health behavior theory to explore the influence of information and experience on arsenic risk representations, policy beliefs, and protective behavior. *Risk Analysis, 26*(2), 353–368.

Sharpe, J. (2009). The mechanism of training educators/teachers in disaster prevention education. *Proceedings, 2nd International Conference on Education for Disaster Prevention.* Douliou, Taiwan.

Shaw, G. L., & Harrald, J. R. (2004). Identification of the core competencies required of executive level business crisis and continuity managers. *Journal of Homeland Security and Emergency Management, 1,* http://www.bepress.com/jhsem

Sheaffer, J. R., Roland, F. J., Davis, G. W., Feldman, T., & Stockdale, J. (1976). *Flood hazard mitigation through safe land use practices.* Report prepared for the Office of Policy Development and Research, U.S. Department of Housing and Urban Development. Chicago: Kiefer & Associates.

Sheeran, P., Trafimow, D., Finlay, K. A., & Norman, P. (2002). Evidence that the type of person affects the strength of the perceived behavioural control-intention relationship. *British Journal of Social Psychology, 41*(2), 253–270.

Shinn, M., & Toohey, S. M. (2003). Community contexts of human welfare. *Annual Review of Psychology, 54,* 427–459.

Siegrist, M., & Cvetkovich, G. (2000). Perception of hazards: The role of social trust and knowledge. *Risk Analysis, 20*(5), 713–719.

Siegrist, M., & Gutscher, H. (2006). Flooding risks: A comparison of lay people's perceptions and expert's assessments in Switzerland. *Risk Analysis, 26*(4), 971–979.

Siegrist, M., & Gutscher, H. (2008). Natural hazards and motivation for mitigation behavior: People cannot predict the affect evoked by a severe flood. *Risk Analysis, 28*(3), 771–778.

Simon, H. A. (1955). A behavioural model of rational choice. *Quarterly Journal of Economics, 69,* 99–118.

Simpson-Housley, P., & Bradshaw, P. (1978). Personality and the perception of earthquake hazard. *Australian Geographical Studies, 16,* 65–77.

Sims, J. H., & Baumann, D. D. (1972). The tornado threat: Coping styles of the North and South. *Science, 176,* 1386–1392.

Sjöberg, L. (1979). Strength of belief and risk. *Policy Sciences, 11,* 39–57.

Sjöberg, L. (2006). The distortion of beliefs in the face of uncertainty. *International Journal of Management and Decision Making, 8*(1), 1–29.

Sloman, S. A. (1996). The empirical case for two systems of reasoning. *Psychological Bulletin, 119*(1), 3–22.

Slovic, P. (1986). Informing and educating the public about risk. *Risk Analysis, 6*(4), 403–415.

Slovic, P., Finucane, M. L., Peters, E., & MacGregor, D. G. (2002). Risk as analysis and risk as feelings: Some thoughts about affect, reason, risk and rationality. *Risk Analysis, 24*(2), 311–322.

Slovic, P., Fischhoff, B., & Lichtenstein, S. (1982). Facts versus fears: Understanding perceived risk. In D. Kahneman, P. Slovic, & A. Tversky (Eds.), *Judgment under uncertainty: Heuristic and biases* (pp. 463–492). Cambridge: Cambridge University Press.

Slovic, P., Fischhoff, B., Lichtenstein, S., Corrigan, B., & Combs, B. (2000). Preference for insuring against probable small losses: Insurance implications. In P. Slovic (Ed.), *The perception of risk* (pp. 51–72). London: Earthscan.

Slovic, P., Finucane, M., Peters, E., & MacGregor, D. G. (2002). The affect heuristic. In T. Gilovich, D. Griffin, & D. Kahneman (Eds.), *Heuristics and biases: The psychology of intuitive judgment* (pp. 397–420). Cambridge: Cambridge University Press.

Smith, K. (1993). *Environmental hazards: Assessing risk and reducing disaster.* London: Routledge.

Smith, J. R., & Terry, D. J. (2003). Attitude-behaviour consistency: The role of group norms, attitude accessibility, and mode of behavioural decision-making. *European Journal of Social Psychology, 33*(5), 591–608.

Speer, P. W., & Peterson, N. A. (2000). Psychometric properties of an empowerment scale: Testing cognitive, emotional and behavioural domains. *Social Work Research, 24,* 109–118.

Spittal, M. J., McClure, J., Siegert, R. J., & Walkey F. H. (2005). Optimistic bias in relation to preparedness for earthquakes. *Australasian Journal of Disaster and Trauma Studies, 1.* http://www.massey.ac.nz/~trauma/issues/2005-1/spittal.htm

Spittal, M. J., McClure, J., Walkey, F. H., & Siegert, R. J. (2008). Psychological predictors of earthquake preparation. *Environment and Behavior, 40,* 798–817.

Spittal, M. J., Walkey, F. H., McClure, J., Siegert, R. J., & Ballantyne, K. E. (2006). The earthquake readiness scale: The development of a valid and reliable unifactorial measure. *Natural Hazards, 39,* 15–29.

Sprott, D. E., Hardesty, D. M., & Miyazaki, A. D. (1998). Disclosure of odds information: An experimental investigation of odds format and numeric complexity. *Journal of Public Policy and Marketing, 17,* 11–23.

Stoll-Kleemann, S., O'Riordan, T., & Jaeger, C. C. (2001). The psychology of denial concerning climate mitigation measures: Evidence from Swiss focus groups. *Global Environmental Change: Human and Policy Dimensions, 11,* 107–117.

Stone, E. R., Yates, J. F., & Parker, A. M. (1994). Risk communication: Absolute versus relative expressions of low-probability risks. *Organizational Behavior and Human Decision Processes, 60,* 387–403.

Strickland, B. R. (1989). Internal-external control expectancies: From contingency to creativity. *American Psychologist, 44,* 1–12.

Tanaka, K. (2005). The impact of disaster education on public preparation and mitigation for earthquakes: A cross-country comparison between Fukui, Japan and the San Francisco Bay area, California, USA. *Applied Geography, 25,* 201–225.

Tatsuki, S. (2000). The Kobe earthquake and the renaissance of volunteerism in Japan. *Kwansei Gakuin University Department of Sociology Studies, 87,* 185–196.

Taylor, S. E. (1983). Adjustment to threatening events: A theory of cognitive adaptation. *American Psychologist, 38,* 1161–1173.

Taylor, S. E., & Fiske, S. T. (1978). Salience, attention, and attribution: Top of the head phenomena. In L. Berkowitz (Ed.), *Advances in experimental social psychology* (pp. 249–288). New York: Academic Press.

Trope, Y., & Liberman, N. (2003). Temporal construal. *Psychological Review, 110,* 403–421.

Thomalla, F., Downing, T., Spanger-Siegfried, E., Han, G., & Rockström, J. (2006). Reducing hazard vulnerability: Towards a common approach between disaster risk reduction and climate adaptation. *Disasters, 30*(1), 39–48.

Thomalla, F., & Schmuck, H. (2004). 'We all knew that a cyclone was coming': Disaster preparedness and the cyclone of 1999 in Orissa, India. *Disasters, 28*(4), 373–387.

Thomas, D. S. K., & Mitchell, J. T. (2001). Which are the most hazardous states. In S. L. Cutter (Ed.), *American hazard scapes: The regionalization of hazards and disasters* (pp. 115–155). Washington, DC: Joseph Henry Press.

Tierney, K. J. (1999). Toward a critical sociology of risk. *Sociological Forum, 14*(2), 215–242.

Tierney, K. J., Lindell, M. K., & Perry, R. W. (Eds.). (2001). *Facing the unexpected: Disaster preparedness and response in the United States.* Washington DC: Joseph Henry Press.

Tönnies, F., Harris, J., & Hollis, M. (2001). *Community and civil society.* Cambridge: Cambridge University Press.

Trafimow, D., & Fishbein, M. (1994). The moderating effect of behaviour type on the subjective norm-behaviour relationship. *The Journal of Social Psychology, 134,* 755–761.

Trewin, D. (2006). *Aspects of social capital: Australia* (No. ABS Catalogue No. 4911.0). Canberra, Australia: Australian Bureau of Statistics.

Triandis, H.C. (1995). *Individualism and collectivism.* Boulder, CO: Westview.

Turner, R. H., Nigg, J. M., & Paz, D. H. (1986). *Waiting for disaster: Earthquake watch in California.* Los Angeles: University of California Press.

Tversky, A., & Kahneman, D. (1981). The framing of decisions and the psychology of choice. *Science, 211*(4481), 453–458.

Tversky, A., & Kahneman, D. (1982. Judgments under uncertainty: Heuristics and biases. In D. Kahneman, P. Slovic, & A. Tversky (Eds.), *Judgement under uncertainty: Heuristics and biases* (pp. 3–20).Cambridge: Cambridge University Press.

U.S. National Research Council (1989). *Improving risk communication.* Washington, DC: PRESS.

Vieno, A., Santinello, M., Pastore, M., & Perkins, D. D. (2007). Social support, sense of community in school, and self-efficacy as resources during early adolescence: An integrative model. *American Journal of Community Psychology, 39*(1–2), 177–190.

Vogt, C. A., Winter, G., & Fried, J. S. (2005). Predicting homeowners' approval of fuel management at the wildland-urban interface using the theory of reasoned action. *Society and Natural Resources, 18,* 337–354.

Völker, B., Flap, H., & Lindenberg, S. (2007). When are neighbourhoods communities? Community in Dutch neighbourhoods. *European Sociological Review, 23*(1), 99–114.

Webb, G, R., Tierney, K, J., & Dahlhamer, J, M. (2000). Businesses and disasters: Empirical patterns and unanswered questions. *Natural Hazards Review, 1,* 83–90.

Weber, E. U. (2006). Experience-based and description-based perceptions of long-term risk: Why global warming does not scare us (yet). *Climatic Change, 77,* 103–120.

Weber, E. U., Johnson, E. J., Milch, K., Chang, H., Brodscholl, J., & Goldstein, D. (2007). Asymmetric discounting in intertemporal choice: A query theory account. *Psychological Science, 18,* 516–523.

Weinstein, N. D. (1980). Unrealistic optimism about future life events. *Journal of Personality and Social Psychology, 39*(5), 806–820.

Weinstein, N. D. (1989). Effects of personal experience on self-protective behavior. *Psychological Bulletin, 105*(1), 31–50.

Weinstein, N. D., Lyon, J. E., Rothman, A. J., & Cuite, C. L. (2000). Preoccupation and affect as predictors of protective action following natural disaster. *British Journal of Health Psychology, 5,* 351–363.

Werner, F., & Scholz, R. W. (2002). Ambiguities in decision-oriented life cycle inventories: The role of mental models. *International Journal of Life Cycle Assessment, 7*(6), 330–338.

Wiedemann, A. U., Schüz, B., Sniehotta, F., Scholz, U., & Schwarzer, R. (2009). Disentangling the relation between intentions, planning, and behaviour: A moderated mediation analysis. *Psychology & Health, 24*(1), 67 – 79.

Winter, G., Vogt, C. A., & McCaffrey, S. (2004). Examining social trust in fuels management strategies. *Journal of Forestry, 102,* 8–15.

Worchel, P. (1979). Trust and distrust. In W. G. Austin & S. Worchel (Eds.), *Social psychology of intergroup relations* (pp. 174–187). Monterey, CA: Brooks-Cole.

Yoshida, K., & Deyle, R. E. (2005). Determinants of small business hazard mitigation. *Natural Hazards Review, 6,* 1–12.

Zaccaro, S. J., Blair, V., Peterson, C., & Zazanis, M. (1995). Collective efficacy. In J. E. Maddux (Ed.), *Self efficacy, adaptation, and adjustment: Theory, research, and application* (pp. 305–328). New York: Plenum Press.

Zaksek, M., & Arvai, J. L. (2004). Toward improved communication about wildfire: Mental models research to identify information needs for natural resource management. *Risk Analysis, 24*(6), 1503–1514.

INDEX